Studying Religion

Studying Religion

An Introduction

Russel T. McCutcheon

Routledge
Taylor & Francis Group

LONDON AND NEW YORK

First Published 2007 by Equinox Publishing Ltd, an imprint of Acumen

Published 2014 by Routledge
2 Park Square, Milton Park, Abingdon, Oxon OX14 4RN
711 Third Avenue, New York, NY 10017, USA

Routledge is an imprint of the Taylor and Francis Group,
an informa business

ISBN: 978-1-84553-011-2   (hardback)
ISBN: 978-1-84553-012-9   (paperback)

British Library Cataloguing-in-Publication Data
A catalogue record for this book is available from the British Library.

Typeset by S.J.I. Services, New Delhi

For Ingrid (1946–2006)

There is nothing more difficult to convey than reality in all its ordinariness.... Sociologists run into this problem all the time: How can we make the ordinary extraordinary and evoke ordinariness in such a way that people will see just how extra-ordinary it is?

—Pierre Bourdieu, *On Television* (1998: 21)

Contents

Preface

Although this short book can surely be used as the basis for an introductory course in the academic study of religion, it certainly is not written as a textbook. Instead, its first eight chapters are written as a readable, general primer comprised of brief chapters across which an argument develops. It is my hope that these chapters will be read by a wide audience, including, but not limited to, undergraduate students. If an instructor decides to use this book in an introductory course, then these chapters (and the supporting material that follows them) sketch a way of conceiving of the material; because the theme of definition is front and center, it provides an opening for instructors to introduce their own descriptive, comparative, and explanatory information, as well as the chance for them to direct their students to the original writings of some of the scholars whose work is discussed throughout. So, if used as the basis for an entire course, my goal is for a type of collaboration to take place, involving an author such as myself who merely plays the role of table-setter, and instructors who bring on the meal in as many courses as their students have the time and inclination to eat.

But as I said, I did not just have the introductory course in mind when writing this book. Because few Departments of Religious Studies can afford to limit their upper-level course enrollments by listing a variety of prerequisites (that is, although prerequisites ensure that students possess a certain level of knowledge and expertise, they can just as easily be the kiss of death insomuch as they limit course enrollment – enrollment being the coin of the modern academic realm), many of the more specialized courses in our field retain the air of an introduction since some of our students find their way into these classes without first going through the lower-level course, where they would have been introduced to what it is we do when we're doing this thing we call the academic study of religion. So, to assist professors who continually find themselves reinventing the wheel despite teaching upper-level courses, I hope that almost any course taught in the field could use this little book, at least for the first week or two, to ensure that all students are up to speed. Although reading such books as Eric Sharpe's classic *Comparative Religion: A History* (cited in the Resources section of this volume) is well worthwhile (as was recommended to all new graduate students when I began my own studies at the University of Toronto), my hope is that this book provides a better place to start for those wishing simply to orient themselves to the field – whether undergraduate or graduate.

Finally, if picked up by general readers prowling the shelves of their favorite bookstore or cruising an Internet book vendor's site, then it is my hope that the volume will provide them with a brief introduction to one area of the human sciences – introducing them not only to the tools and concepts employed but also to some of the more notable past and current scholars who use them.

A few more prefatory comments are in order; a quick look at the table of contents will make it evident that, unlike many other introductory books, this volume has no descriptive survey of what people commonly refer to as the world's religions (though a few are mentioned in the Glossary section). For the purposes of the typical introductory class, I have found it far more interesting to explore how it is that scholars came to the assumption that there is such a thing as, say, Hinduism, rather than taking its existence for granted and then busily describing all of its various components (i.e., its history, myths, doctrines, rituals, social institutions, symbols, etc.). It is my contention that students will remember and use the skills explored in this book far more than they will recite, at some distant point in the future, a number of esoteric details about this or that religion (the product of the 'Who crossed which river and when?' approach to teaching). For, should students become further interested in, say, some groups' dietary codes or their initiation rites, then further detailed study awaits them – studies that are far beyond the scope of an introduction to our field. To rephrase, I believe it is a disservice to students to use such courses to introduce them to the innumerable, momorizable details without first introducing them to the tools used to come up with the details in the first place. And the following volume is just that, an introduction for newcomers, written with the assumption that it is the tools that scholars use, rather than the details, which are most important to consider; for it does not seem to me that the world naturally divides itself into the parts that we assume it to have. Such division – that is, the act of defining or demarcating this from that in order to come to know something about the world – is an activity that we often overlook when teaching students. There is therefore a subtle sleight of hand – often not caught by the undergraduate student – at work when instructors simply assert that this or that person, place, river, action, or book is important and its name and dates therefore ought to be remembered and repeated back on the test. Following a recommendation made by Jonathan Z. Smith in a short essay on pedagogy that once floated out there in cyberspace – but which appears here, published for the first time, as an Afterword that will hopefully prompt readers to reconsider how it is that we come to know, and teach things about, the world – this book attempts to pull up the scholar's sleeves enough to make it plain to the novice reader how we in the academy talk about religion.

Before turning readers loose on the text, I should say a final word or two about the book's structure and how I can imagine it being used. The introduction and eight short chapters, which can be read on their own while ignoring the rest of the book, are followed by cross-referenced discussions of many of the technical terms used throughout the book. In addition, descriptions of the work of a number of the scholars who are mentioned in the main text, along with sample quotations from their work, are included after the Glossary section. Because both the technical terms that appear in the Glossary section, as well as the names of the scholars whose works are summarized in the final section of the book, are **bolded** when they are first used in each chapter, some readers may flip back and forth, getting close to using the book as a printed hypertext. A short bibliography is included for works that are directly quoted in the book's eight main chapters; otherwise, I have forgone citations for quoted material elsewhere in the book, assuming that chasing down quotations will come much later than this brief introduction. Finally, the book ends with suggestions of current general resources in the field for readers interested in pursuing their studies further.

This brief primer is therefore intended as a concise, readable introduction to the hard work of re-conceptualizing this thing many of us commonly call religion, seeing it as a form of ordinary behavior that is comparable to a host of other all too familiar things that we do. That's what I hope to have communicated in the quotation from the late French sociologist, Pierre Bourdieu, that I have used to open the book.

Acknowledgements

Having taught a number of introductory courses since first beginning full-time work as a university professor in 1993, I have many undergraduate students to thank for their patience with a professor who experimented with various ways of making this material worth thinking about – which went so far as once to encourage graduate assistants teaching Mary Douglas's work on 'soil/dirt' to take a handful of the cookie crumbs, or should I say 'soil', from a clay plant pot and eat it, just to make the point of performing some taboo action in a rather more memorable way. No doubt we've all stood in the classroom's doorway, one foot in and one foot out of the room, to illustrate the ritual state of being between social statuses (the notion of liminality, or being on the limen [Latin for threshold], as in, 'If there's no chewing gum in class, then does that rule apply to me now?'), but I bet that few of us have considered course preparation to involve crushing up a whole pack of Oreos the night before a class. I have no doubt that a number of those students still recall that grad student's little stunt to this day. Although the direction outlined in the following pages is not nearly so theatrical or inventive, if it still proves memorable and useful to those who read this book, then all of those past students are to be thanked for entertaining my invitation for them to think differently about material that they assumed to be familiar, and thus uninteresting, when they first walked into the classroom.

Also, many friends and colleagues have influenced my thinking on the following matters – including, but surely not limited to, those with whom I first worked as a teaching assistant at the University of Toronto and those with whom I've taught at three different public universities in the US. To all of these people I express my thanks. Of course, my version of the introductory undergraduate course – first developed at the University of Tennessee, further tweaked at Southwest Missouri State University (now known as Missouri State) when I realized that teaching a world religions course was rather unrewarding for me and less than intellectually challenging for my students, and then fine tuned at the University of Alabama – ranges far wider than what appears in these pages, but the following at least provides an indication of how I try to set the table for my introductory students, as a way of beginning to persuade them that the varying degrees of folk knowledge that they bring with them to my class does not exhaust what they'll learn. Because a handout that I often use to do just that on the first day of class has also turned out to be of use to a number of people at other schools, all of whom teach their own version of an introductory course, my hope is that this short book (which is merely a greatly expanded version of that handout) will come in handy for them as well.

I also express my thanks to Jonathan Z. Smith for permitting me to include his brief essay on pedagogy, 'The Necessary Lie: Duplicity in the Disciplines'. This essay had previously been available only on a University of Chicago website for graduate students who were teaching for the first time. As far as I can tell, the site was taken down sometime in the summer of 2006. Given how influential Smith has been with regard to my own thinking – both in terms of what I study as well as how it is that I go about my work as a scholar and, just as importantly, a teacher – closing this book with his essay is an important tip of my hat to the figure who lurks in the background of the chapters that precede it. Because I have used this essay in many of my own courses, as a way to make the class and its syllabus objects worthy of the students' collective attention, using it to close this book will likely be very useful to instructors and to readers. My thanks to Jennifer Alfano, an undergraduate senior majoring in both English and Religious Studies who worked in our Department for the 2006–7 academic year, who helped to convert and format the essay for inclusion here.

With students in mind, I should also mention two other student workers at the University of Alabama who deserve to be singled out: Kim Davis and Christine Scott; both are graduates of our BA in Religious Studies, and both are now pursuing their MA degrees (in French Literature and Religious Studies, respectively). They both worked with me during the summer of 2005, when portions of this book were written. They are to be thanked for assisting with researching and drafting earlier versions of some of the definitions and for their assistance with collecting information and drafting portions of the scholarly biographies that appear in the final section of the book. In addition, an abbreviated form of some of the following chapters first appeared that summer on a public website with which they helped a great deal. For all this I owe them a debt of gratitude.

Finally, to Janet Joyce of Equinox Publishers – who, like the editor in Michael Chabon's wonderful 1995 novel about dysfunctional academics, *Wonder Boys*, has awaited this little book for far too long – I express my appreciation not only for her patience but also her continued interest in my work and her sense that publishing work in the academic study of religion is a worthwhile venture. It is my good fortune that she is far more stable than Terry Crabtree, the editor in Chabon's novel.

This book is dedicated to my oldest sister, who always told me that I should write something for non-specialists (or, as she phrased it, people like her) to read. Apart from editing an anthology intended for the class-room, all of my publishing has been directed at a specialized audience. After all, given that most of my work has so far been concerned with the history of this one academic field, I can't imagine many people other than scholars of religion reading it – and even among scholars of religion, few seem interested in examining the shape of the field itself, preferring in-stead to immerse themselves in the study of this or that myth, ritual, or institution. Given that I've not written all that much for the novice, I'm not really sure that this is the book my sister had in mind, but I do know that it's surely closer than anything else I've written. So this one's for Ingrid.

Introduction: What is the Study of Religion?

When we say we're studying religion, what is it in the world of human actions that we're talking about? This is a question scholars of religion must ask themselves, right from the start of their studies. For if scholars, like the people whom they study, presume that their word 'religion' refers to something outside of the world of human actions – something that apparently existed well before, and will long outlast, such actions – then how can one even talk about such things? So just what do scholars mean when they say something is 'religious'?

As a way of offering an answer to this question, imagine the following situation, which is likely so ordinary that it will strike you as uninteresting: You walk into a dark room and fumble for the light switch on the wall; finding it, you casually flick it on as you enter the room – but nothing happens. Surprised that you're still in the dark, you quickly flick it back and forth a few more times, much like people who impatiently push the 'Close Door' button on elevators, as if that'll help. But still, nothing happens. With one hand still on the switch, you peer into the darkness, to where you think the light is on the ceiling. 'The bulb's burnt out', you mutter to yourself, as you wonder if you've got any spare bulbs in the cupboard.

Although scholars are often accused of making simple things overly (and unnecessarily) complicated, I think it worthwhile to consider what is going on in this example of routine, day-to-day behavior – an example so mundane that it might strike us as silly to examine it in greater detail.

Based on countless past experiences of walking into dark rooms, as well as rather rudimentary beliefs about such things as electricity, electrical wiring, and a hunch we have about the average lifetime of filaments inside light bulbs, we routinely infer a relationship between a wall switch and a ceiling bulb – an **inference** that usually matches reality so closely that we never think twice about whether flicking the switch has an effect on the bulb overhead. In fact, I'd hazard a guess and say that the person walking into the dark room does not even consciously *believe* that the switch is connected to the bulb, if by 'believe' we mean that they subscribe to a series of principles or propositions that posit a relationship between the switch on the wall and the light overhead. Instead of seeing their belief about the light as a conclusion reached by means of a systematic set of rational processes, or even a bold conjecture that predicts some future state of affairs based on one's past experiences, we might understand it more as a form of unreflective behavior. Much like walking through a open doorway without first stopping to form a well-grounded hypothesis concerning the likelihood that it is in fact an open physical space through which physical objects might pass, reaching for the light switch in a dark room is more than likely not a conscious, **intentional** activity.

As should be evident from this brief discussion of a failed attempt to illuminate the darkness, reality does not necessarily match our expectations, no matter how reasonable those expectations may seem to us. Birds regularly fly into clean windows, intelligent people walk straight into patio screen doors, and sometimes we're left in the dark when the light bulb burns out. Surprises (and sometimes bruises) result when reality does not match our expectations; learning to become curious about the surprises (and curious about why we were surprised in the first place) is perhaps the first step toward becoming a scholar.

Apart from being a practical illustration of how familiar patterns of human behavior do not always match the way the world seems to operate, there is more that we can draw from this simple example of the light bulb. Consider the conclusion we reached concerning the bulb being burned out. Having changed a number of light bulbs in my time, it seems pretty reasonable to infer that the filament has burned out when it finally dawns on us that repeatedly flicking the switch up and down causes nothing to happen. So I'll unscrew it, see if it is blackened, and without even thinking I'll give it a little shake beside my ear to listen for the sound made by the damaged filament, all to confirm my conclusion. If the light still doesn't work after replacing the bulb, I'll likely hold the new bulb close to my ear and give it a shake as well – after all, bulbs, like eggs, don't always make it home from the store in one piece. If the new bulb seems fine, yet still nothing works, I'll start doing some additional problem solving; I'll hunt for a flashlight and find the circuit breakers; I'll wonder about the switch being broken, or whether the wiring itself is having troubles; maybe I'll try some other switches in the house, in other rooms, to see if the power is off all over the house. Perhaps I should go next door and see if the neighbor also has no power. Should I call the electric company to see what's up? Can I afford an electrician?

In the midst of all this, I think most readers would agree that it is highly unlikely that the person left in the dark will conclude that one or more powerful, invisible **agents** had infiltrated their home's electrical system. Not that this is a silly conclusion to draw – far from it. Instead, given the way many of us usually go about problem-solving, it is a highly curious conclusion to draw – or at least it is curious when people do and when they do not make this rather bold conjecture. For there are indeed times in people's lives when they find it completely sensible to conclude just that – that their daily, practical behaviors have little or no consequence to bring about some desired state of affairs; instead, they turn to a series of specific behaviors intended to effect the world at large through the actions of other agents we could call spirits, ancestors or gods.

But why?

Anyone who has owned an old, temperamental car might know exactly what I'm talking about. Those who have the good fortune of driving a new car probably don't think twice about turning the key and driving away. Much like the presumed connection between the switch and the bulb, they go about their daily lives acting as if there is some **necessary** connection between a turn of a car key and the *vroom* of the engine. But not everyone can afford – literally, afford – to have such confidence in their assumptions about the world. Instead, those of us who drive old cars know all too well that inserting the key into the ignition, and then giving it a turn, does not necessarily result in anything whatsoever. To help bridge the gap between the hoped-for outcome and the unpredictable actual state of affairs, such people sometimes develop little habits, like crossing their fingers before trying the ignition, pumping the gas a specified number of times before trying the key, or treating the car as if it was alive, like an old friend of mine in high school who treated his car like a horse, petting the dashboard and revving the engine while saying 'Whoa' – as if it was a feisty, maybe even cranky, young maverick (or old mare?) with a mind of its own (comprising an instance of what we call **anthropomorphism**.

Of course he knew that the car was not alive, at least not in the way you and I imagine ourselves to be alive and to be agents able to accomplish things in the world. But treating it as if it was, as if petting the dashboard had some connection to the outcome of turning the key and cranking the engine, somehow seemed to help. Whether it helped the car to start or, instead, helped him to deal with the anxiety of never really knowing the outcome of his actions (a common theory concerning the psychological function of **ritual**), is, of course, the question that the curious among us will want to ask and for which they'll try to develop answers.

Those readers interested in getting on with the business of describing the ins and outs of the world's religions are likely a little frustrated by now, wondering what all this has to do with the topic at hand. Why start out by talking about burnt light bulbs, walking into screen doors, and starting-up decrepit old cars, when we could just get on with the business of describing ancient **Hindu myths**, studying **Buddhist rituals**, and learning more about **Jewish** holidays?

If that's what you're after, then this is likely not the book for you; instead, you're recommended to find a world religions website, dictionary or textbook – an easy search since both the web and the book market are flooded with them. Because you'll find what you're after in those resources, there's no need to offer yet another descriptive compilation of the whos, whens, wheres and hows of those things we call religions. Instead, this little book is intended for people who find it curious that, no matter how religious (whatever that word may end up meaning) an auto mechanic is, more than likely he or she first checks the spark plugs and wiggles the wires, and doesn't necessarily start by praying over the car, to get the engine idling the right way (though the car's owner might say a little prayer before the bill for the parts and labor appears).

But more than this, this book is also intended for those who find it curious that some people even name one element of human behavior as '**religion**', in the first place, as if it were somehow identifiably distinct from other elements of daily life (the domain we sometimes call **culture** or **history**. For, prior to describing how, and then developing a **theory** concerning why, people are religious, we need to consider why we ought to collect up and name certain human behaviors *as* religious. Case in point: precisely how do we know that Hinduism, Buddhism and Judaism are things that a scholar of religion ought to study?

For instance, consider a recent case that may be well known to people in the region of the US in which I live and work: several years ago the Chief Justice of the state of Alabama's Supreme Court – the highest judicial authority in the state – used private funds to have a two and a half ton granite monument depicting the Ten Commandments as an open book (also bearing inscribed quotations from a number of widely recognized historical influences on the US legal system) built and then erected one night in the lobby of the state's Supreme Court Offices. Given the long, contested nature of **Church and State** issues in the US, his action, followed by his refusal to have the monument removed, resulted in a series of law suits, none of which Chief Justice Roy Moore won, despite his arguments that he was merely following the state's Constitution, which he, inasmuch as he held the office that he did, had sworn to uphold. In the spring of 2004 he was removed from his office for defying a court order to remove the monument, and, on August 23, 2003, the monument was forcibly removed from the lobby (but it then went on a national tour of the US and Justice Moore went on to mount an unsuccessful bid to unseat the state's current Republican Governor).

Question: is this a religious news story or a political one? Neither? A little bit of both? If so, which part of the story is which? If you were a newspaper editor the answer to this question of classification would have practical ramifications, determining on which page, and in which section, you would run the story. Would you feature it on the front page, amidst the day's most pressing political and economic news, or would you run it on the back pages, among the various ads for local worship services? Your decision could then influence how seriously people took the issue – after all, they likely won't know about it if you bury it on the back pages. And if no one knows about it, then, much as with the proverbial tree falling in the lonely forest, it might as well not even have taken place. Moreover, if you featured it prominently on page one, would it be there because this obviously religious news story had political implications (assuming, perhaps, that religion is a private matter that sometimes makes its way into the public sphere) or because the story was political through and through (and even calling the Judge's motives 'religious' was a tactical move meant to score points with a portion of the voting public)? Depending which of these options you selected, you will have likely taken a stand on a variety of fairly complex questions, such as: Is religion a unique domain, separate from culture? If so, does religion influence culture? Does culture influence – perhaps even cause – religion? Are they separate domains that ought never to interact? Just what is religion?

Sometimes people pray to gods or call upon the ancestors and, sometimes, they just replace the light bulb; sometimes references to powerful, invisible beings is considered a sincere, personal statement of **faith** and, sometimes it is heard as a sly rhetorical move doing otherwise unseen political work. Investigating why one results as opposed to the other is what the academic study of religion is all about. So, if you happen to be curious not just about the descriptive when and who and how and where but also about the explanatory why of these behaviors, along with a curiosity over how and why we use the term 'religion', then studying religion might be for you.

1 What's in a Name?

Readers beware: this opening chapter is not about religion. Come to think of it, despite what many readers might think, neither are all of the other chapters. Instead, they are about some of the issues involved in defining an object of study – whatever that object of study may be. Although in our case it happens to be a collection of beliefs, behaviors, and institutions that many people know by the name 'religion', one would think that insights derived from examining how definition works elsewhere would pay off in our field as well.

Before one can study religion, one had better figure out who uses the word (and who does not) and what we, as scholars, mean by it. Does it refer to some real thing out there in the world? Does it refer to something deep inside the human heart? Or is it just a tool some of us use to name parts of the world that we happen to find curious? Just what does it mean to define something as a something?

To start to offer some answers to these questions, consider the following case...

In March of 1856 Andrew Scott Waugh wrote a letter. Twenty-four years prior to this he had joined the team of British surveyors who were carrying out what was then called the Great Trigonometrical Survey of India. Eventually, he became the Surveyor-General in charge of this massive project. But in 1856, having just completed the survey, he wrote a letter to Captain H. L. Thuillier, Britain's Deputy Surveyor-General of India, who was stationed in the city of Calcutta, on the shores of the Ganges River in northeastern India, near the Bay of Bengal. The contents of Waugh's letter were then communicated to London and eventually reached the desk of not only the Royal Geographical Society, which had been monitoring the progress of the survey, but also the British Secretary for the State of India. In his letter, Waugh addressed what up until that time the surveyors had simply named 'Peak 15', a rather tall mountain in the Himalayas, in the region of the world known as Nepal, bordering what was once known as Tibet to the north, but which is today considered by many (though, of course, not all) as part of the People's Republic of China. Among the work carried out was measuring the depths of all of its valleys and the height of all of its mountains. By 1856 it was clear to Waugh that Peak 15 was indeed the highest mountain as yet discovered in the world and he intended to honor his predecessor, the man who began the Great Trigonometrical Survey in the earliest years of the nineteenth century, by naming Peak 15 after him.

In making his recommendation to rename Peak 15, Waugh acknowledged that 'I was taught by my respected chief and predecessor...to assign to every geographical object its true local or native appellation'. Despite assuring his reader that he had always followed this rule in the past, his letter went on to say that what the surveyors simply named Peak 15 was 'without local name that we can discover, whose native appellation, if it has any, will not likely be ascertained before we are allowed to penetrate into Nepal and to approach close to this stupendous snowy mass... In the meantime the privilege as well as the duty devolves on me to assign to this lofty pinnacle of our globe a name whereby it may be known among geographers and become a household word among civilized nations'. In honor of his onetime chief, the man who began the

Great Trigonometric Survey of India so many years before, Waugh suggested that this 'stupendous snowy mass' be named after Colonel George Everest (1790–1866). He therefore christened it with the French designation 'Mont Everest', a suggestion soon changed by Waugh to 'Mount Everest'. And, as almost anyone can tell you today, the name has stuck.

You may be wondering what this 150-year-old story about the naming of a mountain half a world away has to do with our topic. It certainly is not because I wish to sing the praises of British **colonialists** who surveyed India – though, to be honest, given the fact that they measured the mountain to within 33 feet of modern surveys that place it at 29,035 feet (or 8,850 meters), such praise would not be out of place, especially considering the perils associated with their work. Neither am I citing this example to illustrate how such things as measurements are not as objective and as factual as they might at first appear – though that would be a handy way to use this example, especially taking into account that when we say 'Mount Everest is 29,035 feet tall' we need to know that this measurement is in fact the expression of a relationship between two points, the peak's tip and sea level; what's more, we also need to know whether high tide or low tide is what counts as 'sea level' (for example, back in 1802, when the Great Survey began, the high water mark was used as the standard, though later the mid-point between high and low tides was taken as the baseline for 'sea level'). And I am not citing this example simply because Mount Everest has recently been in the news quite a lot, since May 29, 2003 marked the 50th anniversary of the peak being climbed by the New Zealander, Edmund Hillary, and his Sherpa partner, Tenzing Nargay – though the way in which their 1953 ascent was popularized in some parts of the world certainly has helped to create our modern view of the mountain's mystery and majesty. No, the reason I use this example is to put squarely before readers the problem of naming and the issues involved when we set out to define things. For this story sets the stage rather nicely for us to consider the role classification plays in enabling us to know and act in the world around us – whether we are classifying mountains, vegetables, citizens, cultures, or those things some of us call religions.

Before moving on to examine the issues involved in defining and thereby studying religion, let's re-consider this example of naming a mountain. Although it is pretty difficult to think of Mount Everest as anything but *Mount Everest*, things are always more complicated than they at first appear. (I use the italics on purpose, for Waugh was right: its name has become a household word that communicates awe, challenge, danger, and even triumph – but are these qualities inherently in the mountain or the result of social groups projecting the qualities onto it?) In fact, despite the commonsense assumption that the names we give to things reflect, capture or correspond to some key feature of the things being named (called the **correspondence theory of meaning**), the names we give to things may, instead, tell us more about the namer than they do about the thing being named. For example, despite the name for this mountain now being a household word, the fact that we say it the way we do sets us rather far apart from the people who might have read Waugh's letter back in 1856. For if you had the opportunity to meet the good Colonel for whom the mountain was named, and pronounced his name as we do now, 'Ever-est', you would have been quickly corrected to pronounce it as 'Eve-rest'. Although this seems a rather minor example that can be easily overlooked – as in when some people in the US say 'pop' and others say 'soda' for what both agree to be a carbonated beverage or soft drink – it does bring to light the fact that names are products of social worlds that change over time; despite what we usually think, names are not necessarily neutral and objective labels that are placed on things. 'Mount Ever-est' has become such a widely accepted pronunciation that you'd likely have trouble using 'Mont Eve-rest' to conjure up a sense of awe in a conversation with someone, for more than likely they wouldn't be able to get past what they'd hear as your mispronunciation of the name. Whether or not the way in which Sir George Everest pronounced his own name has any bearing on how we say it today – in other words, is it correct to say that we are 'mispronouncing' it, as if his own pronunciation was a standard against which all others are measured? – is a topic to which we shall have to return when studying the way people understand and talk about their own beliefs and behaviors (what we call the insider/outsider problem – a topic raised in the earlier quotation from Waugh's letter concerning his habit of trying, when possible, to use the 'native appellation' for geographical features).

But it is not just the fact that the pronunciation has changed over the past 150 years – as if the meaning was uniform despite differences in the way it was pronounced. Whether or not Waugh knew it at the time, the south side of the mountain (which is the side that is seen from the modern country of Nepal) had long been known as 'Sagarmatha', meaning 'goddess of the sky'; and in Tibet, on the north side of the mountain, it had long been called 'Chomolungma', meaning 'mother goddess of the universe'. Although of little use to the British surveyors, these two local names were obviously meaningful and useful to those who had long used them, for these classifications functioned in relation to systems of belief and behavior that helped to make the Nepalese and Tibetan social worlds possible. And let's not forget that to Colonel Everest himself the mountain was not known as Mount Everest, at least not while he worked on the survey, but simply 'Himalaya Peak 15' – a seemingly neutral designation but one equally immersed in a complex belief and behavioral system, one that had something to do with the importance of measuring mountains in distant lands in the first place, and naming them by using a modern numbering system derived from both ancient Indian and Arabic cultures. Including these names, along with the various pronunciations of Everest's own name and Waugh's original use of the French 'Mont', we have at least six designations for what some could simply understand as a common, and therefore relatively uninteresting, geological formation that began forming about sixty million years ago when one tectonic plate started moving north at about fifteen centimeters per year, grinding against another plate and thrusting the sediment upward.

Voilà, a mountain is born.

This little episode in the history of naming is just the tip of a rather large topic that, when we are going about our daily business, we simply (perhaps necessarily) overlook. But if we pause and refocus our attention – something scholars generally do – we might start asking some questions that we might otherwise not have asked: Just why does a mountain in Nepal bear the name of a British surveyor? Despite Waugh's seemingly well-intended assurance that they always worked 'to assign to every geographical object its true local or native appellation', what were the British even doing in India in the first place and why were they mapping it from top to bottom? Are maps, like the names we give to things, simply neutral representations that correspond to stable land masses or, like height being measured in relation to sea level, which is itself hardly a stable basis, are they expressions of ever-changing relationships, not only between the namer and the thing being named but between competing namers and their competing names and competing interests? If we opt for the former choice, then we take no notice of the fact that a mountain high in the Himalayas is named for a man born in 1790 in Greenwich, England (pronounced Gren-itch), and we simply continue presuming that Mount Everest couldn't be anything but Mount Everest – after all, if this is the way we approach naming then the name bears some direct relationship to some inner aspect, quality or **essence** in the thing being named. A slight variation to this approach is to hold that objects possess some inner characteristic that is only arbitrarily and loosely linked to the labels that we place on them. After all, whether you are an English speaker and call it a book, *livre* in French or *Buch* in German, you still read it. And, whether you know it as pop or soda (or simply calling all carbonated beverages Coke, as they do in the US south), you still drink it when you're thirsty.

But if we do not think that one can so easily separate name from identity – therefore making it rather difficult for young Romeo simply to, as Juliet requests, 'doff thy name, which is no part of thee' – then we might opt for the latter choice and understand the ways that groups define, classify, name and plot things as the tips of very large social and political icebergs bumping up against each other, grinding away at each other. In this case, the rose's sweet smell and its name are not so easily separated. For example, look no further than the very system we now routinely use to tell time and the system we commonly use to plot longitudes (latitudes are determined by the equator) – these are systems of chronological and spatial classification that both make reference to the city in which George Everest happened to be born. Of course, I am referring to Greenwich Mean Time and the Greenwich Prime Meridian of the World, which means that time is measured either as being ahead or behind the time in the city of Greenwich. In the Central Time Zone of the US, where the University of Alabama is located, we are at GMT – 6 ('minus 6' means 6 hours west, and thus 'behind', Greenwich). Moreover, because Greenwich has been given a longitude of 0 degrees, every point on the globe can be measured in relation to being either west or east of this point – much like the use of sea level for measuring height. So, the link between this city in southwest England and the systems we routinely use to plot our place and time on the globe can be seen as (1) an unquestioned natural fact; (2) a neutral, and possibly even arbitrary, relationship, since we had to use somewhere as the starting point; or (3) evidence of the history of British colonial rule, its unmatched naval supremacy over the past several hundred years, and therefore that nation's political and economic dominance of much of the world for quite some time.

So, what's in a name? Apparently, an awful lot.

As we look deeper into the issue of definition, it gets increasingly difficult to see classification as merely a natural, neutral or innocent activity. Instead, classification seems fraught with interests, agendas and implications. It was just this point that was so nicely illustrated in Christopher Monger's film, 'The Englishman Who Went Up a Hill But Came Down a Mountain' (1995). Starring Hugh Grant and Tara Fitzpatrick, this romantic comedy was about two English government cartographers (map makers) who arrive in a small village in Wales in 1917, and find that the mountain that is so loved by the locals – which they refer to as 'the first mountain in Wales' – fails to meet the 1,000 foot minimum height to be designated on the cartographers' map as a mountain. Wounded local pride, mixed together with a long history of antagonism between Wales and the rest of England, prompts the townsfolk to get out their shovels and wheelbarrows and add a few extra feet to the 'hill', ensuring that upon being re-measured it is designated by the official map-makers as a 'mountain'.

If this is the case – if classifications are not innocent or natural but, instead, intimately linked to groups of people with (sometimes conflicting) interests – then one might be forced to ask whether the thing we call either 'pop' or 'soda' is always 'something to drink' – after all, what if you're not thirsty? Perhaps, then, it might be an irrelevant item in your environment that does not even attract your attention, much less prompt you to attach a name, and thus an identity and value, to it.

The late **Mary Douglas**, the well-known British **anthropologist**, once observed in her classic 1966 study of ritual purity systems, *Purity and Danger*, that the difference between 'dirt' and 'soil' was that the stuff we know as dirt was 'matter out of place'. Her point? The same generic material takes on different meanings, values and identities in relation to different classification systems, each of which puts into practice different sets of interests – which changes from time to time, group to group, and occasion to occasion. The same generic stuff of the world, once mapped into one set of preferences, allows us to experience it as 'soil' (say, when it is in a farmer's field or providing nourishment for a potted plant in your home), whereas mapped into another set prompts us to see it as 'dirt' (say, when it gets on your clothes or falls from the pot onto the carpet). So, as Douglas concludes, the concept of dirt is 'a by-product of a systematic ordering and classification of matter, in so far as ordering involves rejecting inappropriate elements'. Rephrased, we could say that the label 'dirt' does not necessarily correspond to something dirty in that which we know as 'dirt'; instead, she seems to be saying: Show me something classified as dirt and I'll show you a classification system that prompts us to distinguish safe from dangerous, allowable from unallowable, clean from dirty – not because there's something inherently dangerous in the things we know as a danger but, instead, because things classified as dangerous threaten interests that are of relevance to a group.

Classification is therefore a social act.

Question: could we see such classifications as **'religion'**, **'myth'** or **'ritual'** as working in the same way, providing evidence of a larger classification system and set of relationships and preferences? In posing this question I'm getting ahead of myself. For the time being, let's just say that just as 'dirt' and 'soil' are classifications that, depending on the circumstances, interests and the choices of the classifier, we place onto the generic stuff of the world in order to transform it from the undifferentiated background noise of daily life into something significant, something worth paying attention to, so too the difference between 'mountain' and 'hill' may tell us little about some geological formation but, instead, may tell us a great deal about the preferences and interests that inform the competing systems of definition used by various groups of people as they make sense of their worlds so as to go about the business of living in them. After all, the movie about adding some height to that mountain in Wales wasn't really about the mountain, but, instead, was all about the identity of a group of people aiming to be something more significant than they might have first appeared.

Because we seem not to have the luxury of getting away with no classifications whatsoever and experiencing reality 'in the raw' – after all, in order to talk about and relate to something we need to place it on our horizon by giving it a slot in our vocabularies and thus placing it in our minds, in our stories, and in our histories – classification, like cartography and surveying, is hardly an innocent business; instead, it is tied up with issues of power and identity. When we leave the realm of map-making and turn our attention to the study of religion – that thing which many people believe to be concerned with the deepest, most enduring issues of significance and meaning yet to be considered by human beings – the problem of classification and definition might seem, at first, to get even more complicated. But the hope is that we start to see that the thorny issues involved in definition apply to all things that we study – from cultures and literatures to mountains and hills. If so, then considering in detail what is involved in coming up with a definition of religion that is useful to the scholar interested in studying human cultures will have implications of other areas of study as well.

And so, with this hope in mind, we leave behind the lofty heights of not only Sagarmatha, but also Chomolungma, Himalayan Peak 15, Mont Eve-rest, as well as Mount Ever-est to consider some of the issues involved in classification, for only once we classify, or define and thereby name, something as a specific sort of something, such as knowing a part of the world as 'religion', will we be able to get on with the work of studying it.

2 The History of 'Religion'

Making the leap from mountains to cultures, this chapter invites readers to consider not just religion as an aspect of wider cultural practices, but the very fact that we think such things as religions exist – that some of us even use the word 'religion' – itself to be a cultural artifact. We therefore begin by acquainting ourselves with the history of the very concept 'religion', keeping in mind that knowing the history, development, and limitations of our concepts may come in handy when we try to use them to name, organize, and move around within our worlds.

Like all items of **culture**, words and the concepts they are thought to convey have a **history** (such as the classification of, and the various associations and value judgments that we make when we hear, the name 'Mount Everest'); not only spelling and pronunciation but meanings and usages change (sometimes dramatically) over time and place. So too, 'religion', and the assumption that the world is neatly divided between religious and nonreligious spheres (i.e., **Church and State**), can be understood as a product of historical development and not a brute fact of social life. Today, long after the modern usage of the word '**religion**' was first coined, it is no longer obvious how it was understood in the past or how we ought to use it today. In fact, it is not altogether clear that scholars should continue to use it when studying human behavior. After all, just because a group of people use a concept as part of their own way of talking about themselves and the rest of the world does not mean that scholars studying these people must use it as well.

That many of the people a scholar might study certainly talk about their religion, their religious beliefs and their religious institutions should be obvious to anyone; however, that many of the people one might study do *not* talk about their world in this manner whatsoever is equally obvious to anyone who has done even just a little cross-cultural work. Therefore, contrary to other introductions that employ the term 'religion' as if it refers to a universal feature animating those social movements called 'the **world's religions**' – a term first coined in Europe not so long ago (on this, see the work of **Tomoko Masuzawa**) – this introduction will be a little more cautious when it comes to making general claims about all human beings. Instead, it will primarily concern itself with the history and use of the idea of 'religion'.

The English word 'religion' has equivalents in other modern languages, e.g., in Germany the academic study of religion is known as *Religionswissenschaft* and in France it is known as *les sciences religieuses*. In nineteenth-century Britain the academic study of religion was simply called either the Science of Religion or **Comparative Religion**, the latter name emphasizing the cross-cultural nature of its data while the former emphasized the systematic and rational manner in which it was studied; **F. Max Müller**, one of the founders of the field, referred to the newly emerging field as the Science of Religion, following the model of what was then the newly emerging cross-cultural field called the Science of Language, or Philology – what we today might simply term **Linguistics**.

A quick comparison, therefore, reveals that languages influenced by Latin and, later, European culture, possess something equivalent to the term 'religion'. This means that for pre-**colonial** contact cultures, or in those few that today remain unaffected by Europe and North American languages, cultures and economies, there was not a necessarily equivalent term. For instance, consider the case of modern India; although 'religion' is not a traditional concept there (that is, the ancient language of **Sanskrit** long predates the arrival of Latin-based languages in that part of the world that we today call India), the effects of British colonialism ensure that contemporary English-speaking citizens of the modern **nation-state** of India have no difficulty conceiving of what is called '**Hinduism**' as their 'religion' – although, historically speaking, that which world religions textbooks call 'Hinduism' can also just be understood as **sanatana dharma**: a Sanskrit term for the cosmic system of duties and obligations that affects all aspects of *samsara* (which itself names the almost endless cycle of births and rebirths). Although the uses and meanings of *sanatana-dharma* and 'religion' may indeed overlap to some extent, assuming that they are one and the same – that '*sanatana-dharma* means religion' – is likely an unwise move for the careful scholar to make.

Consider another case: even the **Christian** text known as the New Testament is not much help since its language of composition – known as *koiné*, or common, **Greek** – also predated Latin precursors to our modern term; the authors of the many texts that comprise the New Testament therefore lacked the linguistic roots from which we today derive our word 'religion'. So, although English translations routinely use 'religion' or 'godliness' to translate such ancient Greek terms as **eusebia** (as found in 1 Timothy 3:16 and 2 Timothy 3:5), or *threskia* (a term that referred to the ceremonials of worship, as found in Acts 26:5 and James 1:26, 27), these ancient Greek terms are much closer to the Sanskrit *dharma*, the Chinese **li**, and the Latin **pietas** – all words having something to do with the quality one is thought to possess as a result of properly fulfilling sets of social obligations, expectations and **ritual** procedures, not only toward the gods or the ancestors but also to one's family, peers, superiors, servants, etc. For instance, despite 'piety' today meaning an inner sentiment or affectation, to be pious in ancient Athens – what the ancient Greek philosopher Socrates was accused of *not* being, as the story is told in Plato's dialogue on defining *eusebia* (piety), entitled *Euthyphro* (c. 380 BCE) – meant recognizing, and publicly signaling that you recognized, differences in social status and privilege. This, of course, is the great irony of the tale told in the *Euthyphro*: in this tale Socrates's accuser is an impious young upstart, and Euthyphro (who is intent on instructing Socrates) is an outright braggart; by their behavior the ancient reader would have known that neither can judge either *eusebia* or Socrates.

Or consider one final case, that of the Arabic term **din** (pronounced 'deen'), which is today commonly translated as 'religion'. According to the *Encyclopedia of Islam*, the modern term/concept *din* seems to have developed from the much earlier notion of a debt that must be settled. Eventually, the concept develops such that we find the phrase *yawm al-din*, or what we might translate as the 'Day of Judgment', when Allah (from Arabic, translating simply as 'the God') gives direction to all human beings. What we therefore find in *din* is – much like ancient senses of piety – a term that once operated within a world of social exchanges, a world of social rank (as in debts owed), and a world of social rules and obligations. It then moves from the more narrow or mundane sense of a debt to be settled to being a term that stands for the entire collection of required directives to which one must submit – as in one's submission to the will of Allah. Therefore, it could be rather misleading to suggest that '*din* means religion', as many people do – especially if by 'religion' one means what so many people in the English-speaking world do: '**faith**', a unique type of inner **experience**, or 'belief in God'. Instead, what we find in the example of *din* is an Arabic term translated by contemporary English speakers who assume that the word 'religion' has universal significance and therefore must have equivalents elsewhere in the world.

The danger that we as scholars run into when assuming that our terms are universal is that much is lost in the translation. For instance, consider a recent translation of the Qur'an's famous *sura* (or chapter) 5:3: 'Today I have perfected your system of belief and bestowed My favours upon you in full, and have chosen submission (al-Islam) as the creed for you.' Or, as phrased in another English translation: 'The day I have perfected your religion for you and completed My favour to you. I have chosen Islam to be your faith.' Both 'system of belief' and 'creed' in the first translation, and 'religion' and 'faith' in the second, are English renderings of *din* – translations that nicely lock the Arabic social term within a world of private sentiment (that is, belief or faith). Nothing could be further from the complex social, transactional history of the ancient Arabic concept. Moreover, knowing something of *din*'s etymology, or historical development, sheds important light on one's understanding of **Islam**, given that the notion of submission (as in submitting to the will and directives of Allah) plays such a central role. It therefore seems that one should be careful not to conclude too quickly that *din* – not to mention *dharma*, *eusebia* and *pietas* – means religion. Meaning, it appears, is not such a simple matter of looking for identities and **correspondences**.

So where does all of this leave us? Recognizing not only that a word's history holds no clue concerning how we ought to use it today – for, as made clear in the glossary's entry on 'religion', there's nothing religious about the ancient Latin roots of our modern word religion – but also that we do not easily find synonyms in other cultures for the words/concepts that we take for granted, scholars find a number of questions in need of investigation: If a culture does not have the concept, can we study 'their religion'? Should scholarship only employ concepts local to the group under study? If so, can one, for example, study someone's 'culture' or their 'DNA' if they lack the words? Or, despite its local nature, is the thing to which our word 'religion' points shared by all people, regardless of their self-understandings and their vocabulary – much as Shakespeare wrote in 'Romeo and Juliet': 'a rose by any other name would smell as sweet'? But is using our local term as if it were a universal signifier an act of cultural imperialism?

These are important questions for those who attempt to develop a cross-culturally useful definition of the religion concept, distinguishable from its popular or folk definition. After all, just as chemists develop a technical vocabulary, driven by **theories** concerning how chemicals and molecules interact, that prompts them to talk about 'H_2O' instead of merely 'water', or, as the scholar of Christian origins, Willi Braun, has pointed out in the opening chapters to the *Guide to the Study of Religion*, just as astronomers use 'crater' to talk about something others might simply call a 'hole', bringing with their technical term a complex series of theories concerning the movement of physical objects in space, so too scholars of religion who aim to study religion as an aspect of the social world develop technical categories capable of working with cross-cultural data. 'Concepts are not given off by the objects of our interest', writes Braun, '[t]hey neither descend from the sky nor sprout out of the ground for our plucking.' So, as with **anthropologists** who study 'culture' – yet another Latin-based term that is alien to many of the world's 'cultures' – the challenge, then, is to work from the ground up: to take contextually and historically specific words, and the concepts they entail, and retool them for use in studying diverse historical and geographic settings.

3 The Essentials of Religion

As a first step in retooling the concept religion, to make of it a term that might be of use in talking about the world of human actions, we consider what might be the most common approach to defining anything, let alone religion: essentialism, or the approach that assumes an enduring identity, core, substance, or, simply put, essence lurks deep within objects, making them what we say they are.

A notable early attempt to develop a technical – rather than relying on a common or folk – definition of religion as a universal human feature was that of the nineteenth-century **anthropologist, Edward Burnett Tylor** (1832–1917) in his influential book, *Primitive Culture* (1871, 2 vols.; reprinted today as *Religion in Primitive Culture*). A 'rudimentary definition of religion', he wrote, 'seems best to fall back at once on this essential source...belief in Spiritual Beings'. In this classic, minimalist definition we see the still common emphasis on religion as a private, intellectual activity (that is, **religion** equals believing in this or that, as if it is all about what goes on between your two ears) rather than an emphasis on, for example, the behavioral or the social components, as in **Emile Durkheim**'s (1858–1917) emphasis on public **ritual** and social institution in his still influential **sociological** study, *The Elementary Forms of Religious Life* (1912). As stated in Durkheim's often quoted definition: 'religion is a unified system of beliefs and practices relative to sacred things, that is to say, things set apart and forbidden – beliefs and practices which unite into one single moral community called a Church all those who adhere to them... In showing that the idea of religion is inseparable from the idea of a Church, it conveys the notion that religion must be an eminently collective thing.' Unlike Durkheim's sense of religion as something eminently *social* that you do with your body (making it public), for Tylor, religion is an eminently *individual* thing that you do with your mind (making it private).

In Tylor's onetime popular definition we therefore find the remnants of a philosophically **idealist** era in European history, when one's membership within certain groups was thought to be primarily dependent upon whether one *believed in* something (for example, agreed with the claims made in a creed or in a pledge of allegiance), rather than membership being the result of collective behaviors, such as a group of soldiers saluting a flag or people standing in unison to sing a national anthem (as argued in Durkheim's work). In fact, this idealist presumption still persists today, insomuch as the institutions some scholars of religion refer to as 'the cumulative tradition' (the mere externals of religious **experience**, they might say) are thought to be a somewhat deadened (that is, unreflective, automatic, etc.) behavioral expression of a prior, dynamic affectation often known as '**faith**' or 'belief' (e.g., see **Wilfred Cantwell Smith**'s classic 1962 work, *The Meaning and End of Religion*). We can easily find something like this distinction in popular culture today, in which people regularly distinguish between something they call spirituality, on the one hand, and on the other the institution of religion. 'I'm not religious', they say, 'I'm spiritual' – translation: 'I do not participate in unthinking ritual and pointless institution but, instead, participate in an inner, personal quest.' That such people did not come up with this way of thinking on their own, let alone originate the particular path on which they say they are traveling, indicates that they too are part of long established traditions and institutions with rituals of their own – such as saying 'I'm not religious, I'm spiritual'. It's just that they are participating in a *different* and more than likely *competing* tradition, requiring devices to distinguish it and, then, authorize it over the others.

With its emphasis on the intellectual or **cognitive** component (along with such other early scholars as **Herbert Spencer** [1820–1903] and **James G. Frazer** [1854–1941], Tylor is numbered among a group today called the Intellectualists, a nineteenth-century anthropological tradition), Tylor's work offers an example of a classic definitional strategy: **essentialism**. Because the social movements classified as religions struck such observers as obviously having a number of different outward characteristics, many of which were explained away as mere **historical** accidents (i.e., the result of specific cultural or geographic context), they thought it unwise to define religion based on what they took to be its merely secondary, external aspects. Instead, like many others, Tylor reasoned that one ought to identify 'the deeper motive which underlies them'. Belief in spiritual beings, he concluded, was just such a deeper and thus universal motive; in fact, he concluded that it was the 'essential source' for all religions. Accordingly, his **naturalistic** theory of religion sought to account for the universal belief in spiritual beings (a theory known as **Animism**). We therefore refer to Tylor's definition as essentialist (sometimes also termed substantivist or **monothetic**): it identifies the one essential feature (or substance) *without which something would not be what it is.*

In other words, if, as the German Protestant **theologian Rudolf Otto** (1869–1937) once argued in his influential book, *The Idea of the Holy* (1917), that which sets religions apart is the participant's feeling of awe and fascination when in the presence of what Otto termed the ***mysterium tremendum*** (the compelling yet repelling mystery of it all), then without evidence of this sense of awe and fascination there is no religion. This feeling of utter awe (a complex combination of fear, trembling, fascination and attraction) was, for Otto, the essence of religion – something that could only be apprehended by the participant. For the late eighteenth-century German theologian **Friedrich Schleiermacher** the essence was 'a feeling of absolute dependence'; for the early twentieth-century Dutch **phenomenologist** of religion, **Gerardus van der Leeuw**, it was 'power'; for the early twentieth-century theologian **Paul Tillich** it was 'an ultimate concern' (which he once defined, in an unhelpful circular fashion, as 'a concern about the truly ultimate'); for the late nineteenth-century **psychologist** of religion, **William James**, it was an experience peculiar to so-called 'religious geniuses' that, once expressed, taught, reproduced, and, finally, institutionalized, was prone to deteriorate; and for the historian of religions, **Mircea Eliade**, it was the experience of the **sacred** – which he defined as 'not the profane' (a rather less than useful definition, for it begs us now to define **profane**).

Although Tylor's classic definition differs significantly from all of these others (insomuch as his anthropological, or **etic**, perspective aimed first to document and then to explain the cause of such beliefs, whereas the others all presumed that the object of the belief existed independently of believers, thus prompting their responses), all of these scholars went about the task of definition in the same manner: the **inductive** method was used, whereby one compares a number of **empirical** examples, looking for their underlying similarity. We see here the common strategy of employing the comparative method to identify non-empirical commonality, such that certain types of all too obvious difference are understood to be nonessential features of contingent history – an approach characteristic of a number of scholars, from Frazer's multi-volume *The Golden Bough: A Study in Magic and Religion* (1st edition 1890) to Mircea Eliade's (1907–1986) classic work, *Patterns in Comparative Religion* (1949). Today, this approach is most evident in the work of scholars of religion who attempt to identify the deep similarities among the **world's religions** – an effort that generally goes by the name of religious pluralism or **inter-religious dialogue** (e.g., **Martin Marty** and **Diana Eck**). Thus, the study of religion, at least as carried out by some contemporary scholars, is an exercise in identifying what is asserted by some to be a deeply human, and thus humane, element – sometimes called the Human Spirit or **Human Nature**. Based on this presumably shared item, feeling or value, mutual understanding across cultural and historical divides is believed to be possible; after all, studying 'their' sacred symbols, narrative or practices inevitably strikes a chord with 'us' (e.g., the cross-cultural comparative work of **Wendy Doniger**).

Much as with a light switch that can either be on or off – there's no such thing as a light being only partially on, right? – essentialist definitions lead one to name something as religion if, *and only if*, it possesses a certain quality or trait – sometimes called its substance, such that an essentialist definition can also be known as a substantive definition. That just what characterizes this essential quality differs (sometimes dramatically, as evidenced above) from one essentialist to another ought not to be overlooked.

A classic attempt to define something's essence can be found throughout the work of the ancient Greek philosopher, Plato (427–347 **BCE**). Writing in a dialogue style – something akin to reading a play, in which

different characters represent contrary viewpoints, all of which collectively explore a topic, such as 'What is Justice?' – Plato's efforts seem to
have been directed toward identifying that one quality without which
something was not what you said it was. Of particular interest to us is
Plato's already mentioned dialogue, the *Euthyphro*. This dialogue takes
the form of a chance encounter between Plato's teacher – and the main
character in his dialogues – Socrates (470–399 BCE), and a younger man
named Euthyphro, both of whom are about to enter the Athenian law
courts. Socrates is there to be prosecuted for being impious; the charge
against him is that he corrupts the minds of the youth and invents new
gods not condoned by the Athenian city-state (as this charge is laid out in
the dialogue that usually follows the *Euthyphro* in modern editions of it,
entitled the *Apology* [from the Greek *apologia*, meaning to speak in defense of a position]). Euthyphro, we learn, is there to prosecute his own
father for the murder of a farm laborer (or, to be frank, a slave) who was
himself a murderer. Given that charging one's own father with a crime
meant that one risked being judged impious – in fact, Socrates is quite
startled to hear Euthyphro say he was charging his own father with a
crime, for what child knows better than a father? – Socrates proceeds on
the assumption that Euthyphro must indeed know what marks the distinction between a truly pious and an impious act. Otherwise, why would
Euthyphro perform an action that dishonors his own father? Because of
his own impending trial for being impious, Socrates could use an expert's
opinion on what is and what is not pious (or, as we've already seen, what
in ancient Greek was termed the quality of **eusebia**). The short dialogue
that follows, with all its twists and turns, is thus begun with Socrates'
simple and seemingly naive – yet terribly persistent – effort to have
Euthyphro arrive at a clear and defensible definition of piety. (Whether
Socrates is read as a sincere or sarcastic dialogue partner is, of course, left
up to each reader.)

'Tell me, then', asks Socrates, 'what is piety and what is impiety?'

But, before proceeding, it is worth noting, as does Luther H. Martin in
his little book, *Hellenistic Religions* (1987), that what for us might be
understood as 'religion', a seemingly obvious concept most often distinguishable from such other social institutions as politics or economics, the
ancient Greeks might have considered to be divisible into three rather
distinct things: piety (*eusebia*), mystery (Greek *myo*, meaning to shut one's
eyes out of fear or danger) and *gnosis* (from the Greek, meaning secret or
esoteric knowledge; from which we today derive the term **agnostic**).
Whereas the first has already been discussed and will be elaborated below, the second refers to **cults** in which members were initiated into the
mysterious workings of the cosmic order (implying the relations between

mortals and immortals), and the third refers to a tradition in which one's personal salvation was thought to depend upon gaining special, privileged knowledge (as opposed to intellectual or philosophical knowledge; *gnosis* is therefore to be distinguished from another Greek term, *episteme*, meaning rational knowledge, from which we get our philosophical term epistemology, the rational study of knowledge systems). If one is interested in studying 'ancient Greek religion' – especially when using 'religion' in the fashion we've come to use it today as a system of belief in gods and experiences associated with these beliefs – one must therefore collect together beliefs and behaviors from these three otherwise distinct ancient aspects of Greek society, as if they were all somehow essentially interrelated. Not that this is wrong, but it is hardly the way ancient Greeks might have understood the way their world was organized.

Is there something being lost in so easily assuming that our way of understanding the world is shared by everyone?

Perhaps, for as already noted, in ancient Greek, the term *eusebia*, or in Latin **pietas**, signified reverence, honor and esteem – notably as expressed in social and legal relationships. They were not primarily concerned with proper beliefs about invisible powers. Instead, examples would include the proper relations between husband and wife, parent and child, master and slave, soldiers and commanders, and in addition, between mortals and the immortals. Of course a pious Greek would engage in the proper rituals with regard to various gods and ceremonies but would also ensure that relations with social superiors, peers and inferiors were carried out according to the rules of propriety. In Plato's ancient Greek society, *eusebia* therefore signified a wide system of ordinary social practices concerning one's relations with many different sorts of others – relations that extend from the family to the gods, from social inferiors to superiors.

Because we are told that philosophy, for Socrates, was the effort to obtain self-knowledge in order to live the just and worthy life (as he is said to have remarked, 'The unexamined life is not worth living'), then critical inquiry into one's actions and motivations was considered the basis for philosophical inquiry. Because one's actions are rarely, if ever, completely private, but instead are social in origin and implication, we can see not only the social role played by Socratic philosophical inquiry (e.g., the dialectical, or back-and-forth, question/answer teaching method) but, more importantly for us, the intersection between philosophical inquiry and claims to piety. In other words, since piety necessitates knowing, and then acting on, the right social relations, philosophy (that is, public argumentation) can scrutinize those who claim to possess the quality of piety. Such scrutiny is a critical activity since it examines claims that present themselves as beyond examination (somewhat akin to the academic study of religion, perhaps?). For example, it can demonstrate that to boast to have knowledge of piety (as does Euthyphro in Plato's dialogue) is to lack self-knowledge; such a boast would then, ironically, constitute an act of impiety.

In the *Euthyphro* we have a number of such ironies that no doubt would have been immediately evident to the ancient Greek reader: we have a young man who boasts of having privileged knowledge of what is and what is not pious, and he is bold enough not only to charge his own father with a crime but also to set about instructing an older man renowned for his own skills at critical inquiry (i.e., both his father and Socrates are Euthyphro's social superiors). And Socrates has been charged with being impious by yet another young man (who, we're told, can't even grow a decent beard). Issues that involve transgressing rank, honor and privilege should immediately come to the reader's mind. (Speaking of transgression, perhaps **Mary Douglas**'s insights into classification and taboo have some relevance here?) What is therefore of interest is that a dialogue attempting to define the essence of *eusebia* is set within a context ripe with conflicts over rank and power. For the sake of justice, Euthyphro believes that he must prosecute his father – in spite of the fact that the man is his father – rather than exercise the 'proper' and conventional relations (entailing respect and esteem for one's father). Socrates therefore seems justified in inquiring if Euthyphro really is aware of just what constitutes piety. So Plato's character Socrates sets about practicing his **dialectical** method on Euthyphro, the very method of inquiry that has landed Socrates in a lawsuit of his own.

To Socrates' seemingly straightforward question, 'Tell me, then, what is piety and what is impiety?', Euthyphro offers a variety of answers, each of which Socrates critiques on various grounds, until Euthyphro is led to repeating one of his previous yet inadequate definitions. The various definitions offered by Euthyphro are as follows:

1. Piety means prosecuting the unjust individual; impiety is not to prosecute.
2. Piety is what is pleasing to the gods; impiety is displeasing to the gods.
3. Piety is what all the gods love; impiety is what they all hate.
4. Piety is service to the gods; impiety is no attention to the gods.
5. Piety is to know that one's words and actions are acceptable to the gods.
6. Piety is the science of asking of the gods and giving to them.
7. Piety is the art of carrying on business between gods and human beings.
8. Piety is that which is loved by the gods (see 3 above).

When people are asked to define something, they often start by providing an example of it. In the case of Euthyphro, to Socrates' seemingly innocent question, 'What is piety?', he first answers that prosecuting criminals is pious (or 'Doing what I am doing now', as phrased by Eurthyphro in a line from the dialogue that is difficult to read as anything but pompous). But Socrates is unsatisfied, for this, he answers, is not a definition of piety. If you were asked to define, for instance, 'tree', would it suffice for you to answer, 'Yes, a maple is a tree, so is an oak, and an aspen, and a redwood, and also a willow'? There is a difference between offering a definition and providing an example, no? So it is reasonable that someone might inquire about the criterion (or criteria, if more than one) that you employed to narrow down the many things of daily experience to just these five items. That criterion is part of your essentialist definition. Socrates is interested in that one thing that all pious actions share – their essence – for, if he can identify what it is about prosecuting criminals that makes it pious, then he can employ that criterion to classify other pious acts that share this trait – something that would be of tremendous use in his own defense. (Aside: do you see how definition and utility, or usefulness, are linked? Socrates seeks to define piety because he needs to defend himself. Thus, could we hazard a guess that, instead of being right or wrong, accurate or inaccurate, definitions are more or less useful, all depending on one's goals? Could it therefore be that definitions are tactical tools that allow one to get on with producing knowledge about the world – a knowledge that, instead of being innocent and objective, is intimately linked to social interests?)

For our purposes, the key to this dialogue is Socrates' seemingly simple question that follows Euthyphro's third attempt at a definition: 'Is something pious because the gods love it or do they love it because it is pious?' In other words, what makes mere stuff (like Mary Douglas's 'matter' that can either be classified as 'soil' or 'dirt', depending on what you want to do with it [interesting how definition, utility and interests have arisen yet again]) into specific things to which we give names and values: is it our actions toward them that give them value and significance (e.g., loving them, despising them, needing them, ignoring them, etc.) or do we, instead, passively recognize in them some inner quality or substance, beneath their surface – some essence that prompts us to act accordingly toward them? Case in point: is beauty, as some people say, 'in the eye of the beholder' (meaning that beauty is a function of each person's taste or aesthetic sense) or is it more than just skin deep? Does beauty reside deep within the object of art itself, awaiting our careful recognition? As applied to 'religion', is there some inner essence shared by all things that are religious, or is 'religion' a classification some of us give to certain human actions, actions that we select from a wide array, giving them a label we use for a purpose and these purposes change from person to person? Does religion have an essence or is it a function of human behavior, needs and interests?

Knowing that by the end of Plato's text Euthyphro quite literally runs off (either in frustration or embarrassment), leaving Socrates' quest for the essence of *eusebia* unfinished, suggests that, despite Plato's apparent preference for an essentialist definition (evident throughout his many dialogues), the search for an essence may be a hopeless affair. For, as already indicated, there may be as many essences as there are essentialists – or, to put it another way, there may be as many essences as there are interests that drive the effort to nail down the world as if it were comprised of stable, uniform pieces. So, perhaps we ought to pay attention to these needs that drive our efforts to define and make generic aspects of the world into items of **discourse**. If so, then it is to the **function** of religion that we should turn when seeking to define it in a manner that will be of use to scholars of religion interested in studying religion as an item of human history and culture.

4 The Functions of Religion

Because of difficulties with essentialist approaches to definition, many opt instead to define objects not by what they are said to *be* but by what they are observed to *do*. Objects do not therefore have an essence, such scholars argue, but they do have a role to play – they have a function and a purpose – and that is something we can see and therefore study.

With the **essentialist** approach in mind – an approach adopted by those who presume that religions house a core **experience** or fundamental trait that sets them apart from all other aspects of human behavior – we can contrast it with the **functionalist** approach. Consider the thing that appears in many classrooms: a lectern behind which the professor stands while teaching. What is the difference between, say, a lectern and a pulpit? Or, to put it another way, how do we know which name to give to which object and how does the name that we give to it affect how we relate to and value it? Is there some key feature that we can recognize to distinguish between the two, such as color, height, weight, or the material out of which each is made? This does not seem likely, because the same physical object could just as easily be identified as both, not to mention the countless differences among those things that get to count as either a lectern or a pulpit. So what makes a pulpit a pulpit, and not a lectern (not to mention it not being a podium)?

For the functionalist scholar, there is no specific, essential feature that unites all things we call 'lecterns' and thereby distinguishes them from those things known as 'pulpits.' Instead, the context in which something is found, the expectations placed upon it by its users, and, most importantly perhaps, the purpose it serves, are what cause things to be defined as a this and not a that. Functionalists, then, are people interested in asking what something *does* rather than what it *is*.

For early twentieth-century scholars, it was this shift from, as they might have phrased it, *speculating* on universal, non-empirical qualities and affectations to *observing* the role of local, historical context and **empirical** effects that signified the development of what they considered to be a truly scientific (i.e., rational, historical, documentable, testable, etc.) study of religion, in distinction from a well-meaning but, nonetheless, **theologically**-motivated study of religion's enduring value or groundless speculations on its pre-historic origins and **evolutionary** development. For example, consider that group of scholars already mentioned, a group which predated the rise of functionalism: the Intellectualists (known by this name due to their presumption that all human beings, past and present, shared certain intellectual traits). This group of nineteenth-century anthropologists (or perhaps we should refer to them as the precursors to the field today known as **anthropology**) were very interested in origins – to explain something, they assumed, required one to account for its original state and, then, its change over time. Let's consider **Tylor**'s theory of **Animism**, which we have already mentioned; to explain why people today believe in spiritual beings Tylor performed what we might call a thought experiment. He asked his readers to imagine an early human being waking from what you and I commonly refer to as a dream. However, unlike modern people, who have a fairly complex understanding of the difference between being awake and asleep, this 'savage philosopher,' as Tylor termed him, was (he, like many of his contemporaries, assumed) rather child-like and therefore not aware of just what a dream was, let alone possessed with what some today consider a common sense assumption about dreams being the site where subconscious needs are expressed symbolically (a widespread assumption that we owe to the enduring influence of **Sigmund Freud**'s work). Despite the very real differences that likely existed between us and our ancient relatives, Tylor argued that his hypothetical early humans were still much like us for they were profoundly interested in accounting for unanticipated things in their environment (what we might refer to as anomalies). Like us, they were curious problem solvers. It's just that they were evolutionarily earlier than us and did not have access to the sort of scientific methods and intellectual capabilities that we do. Nonetheless, using the rudimentary methods and skills that were at hand, Tylor concludes that these savage philosophers must have arrived at a satisfactory explanation to account for the odd thing of experiencing themselves elsewhere or why they were able to interact with long dead ancestors (all of which you and I know to have been taking place in their dreams, of course). Concluding that there must be something that can leave the body, even outlive the body – perhaps we can simply call it a soul? – struck Tylor as a pretty sensible way for such an early human being to solve this puzzle and account for such an odd experience. Therefore, for Tylor, the **natural** cause of religion – the belief in spiritual beings, or, as he called it, animism – could be explained by understanding its origin, and its origin was based on our ancient ancestors solving puzzles by using their inevitably faulty reasoning about how the natural world worked.

Of course, few today would classify themselves as Intellectualists in quite this way (though the presumption of common **cognitive** mechanisms among human beings is now shared by many scholars), for there are a number of problems with this form of scholarship. To name but one, consider Tylor's conclusion: the 'savage philosopher' explained his experience as the result of his having a soul. As appealing as this theory may be to some readers, we should ask a simple question: is Tylor's theory of animism correct? To answer this, we need to determine a set of criteria so that we can judge whether Tylor's theory of animism is persuasive and accurate. But here we run into a brick wall for, short of inventing time travel to allow us to be present when his early human awoke, so that we could observe him and ask him questions (but in what language?), we cannot really come up with a way to *test* Tylor's theory, to demonstrate it to be either right *or* wrong (something like listening to the bulb while shaking it to test our hypothesis about it being burnt out). It is for this reason that some came to question whether Tylor even had a **theory**, for theories are, by definition, empirically testable (thus, for example, Intelligent Design, the view that a cosmic designer created the universe, is, for scientists [especially those who identify with the **positivist** tradition], not a theory whatsoever), as opposed to an untestable hunch, guess or speculation, premised on the assumption that early humans must have behaved in a childlike manner, solving problems as do contemporary children (after all, nineteenth-century scholars often referred to early humanity as 'the childhood of the species'). To make a long story short, the problem with explaining contemporary events in terms of their origins is that the origin is long gone and remains only as a product of **modern** speculations that cannot help but to project contemporary assumptions backward in time, much as with Tylor's rather Victorian sense of what early humans must have been like (crude but dogged problem solvers). As such, one guess *might* be as good as any other because the standard against which one measures the guesses – the actual origin – cannot be retrieved.

Given just how troublesome it is to try to explain current events in light of their origins – such as answering a question concerning why people sometimes throw salt over their shoulders by saying, 'Well, a long time ago...' – one can see the appeal for some late nineteenth-century, and many early twentieth-century, scholars to shift the ground considerably. Instead of speculating on a long lost original essence or state of affairs, they tried instead to account for the contemporary, observable purpose something serves or the need it fulfills. Functionalists – who no doubt benefit greatly from earlier generations of scholars intent on explaining religion by means of appeals to its historical origins (such as **David Hume**'s work) – made just this switch, seeing their Intellectualist predecessors as doing something other than science. This new focus on the contemporary and the observable made functionalist theories testable, and, as they argued, truly scientific.

Today, functionalists who study religion owe much to three writers in particular, each of whom helped to establish the modern sense of three different approaches in the **human sciences**: political theory, social theory and psychological theory.

Karl Marx (1818–1883), whose **materialist, political economy** theorized religion as a social pacifier that both deadened the oppressed people's sense of pain and alienation while simultaneously preventing them from doing something about their lot in life since ultimate responsibility was thought by them to reside with a being who existed outside history and who would compensate for this-worldly suffering and exploitation in the life to come. Religion, for a Marxist scholar, therefore functions to reproduce the status quo by distracting attention from the actual source of conflict. A quick analogy might help: much as buying lottery tickets uses up financial resources that might otherwise have been used to help one get out of a situation that requires such desperate gambles as buying lottery tickets that are most likely losers, so too focusing energy and resources on religious rituals aimed at one's status in the next life only provides a momentary diversion, much as with the hope that comes along with buying a lottery ticket – until the winning numbers are announced, that is. For Marx, there were better places to invest one's energies than religion – working to change those political and economic conditions that required many people to hope for justice in the next life, rather than this one, was one such place.

Emile Durkheim (1858–1917), whose **sociological** study of religion has already been mentioned, understood intertwined sets of beliefs and practices to enable individuals to form the idea of a common social identity; unbeknownst to participants themselves, for Durkheim the claims of religion were actually symbolically coded claims about the social group itself, since the practices we call religious are none other than members' efforts to assemble and experience the group as an empirical reality (for example, through performing common **rituals** within sight of each other, group members visually experience a group that, normally, exists only in their minds as a shared idea). The collection of narratives, actions and institutions that we call religion, for a social theorist in Durkheim's tradition, therefore functions to build and retain a group identity that is always on the brink of breaking down. (If you do not agree that social identities are fragile things, just think for a moment about how important it is to mail that birthday card to a relative, make that phone call to a parent, or make the proper sort of eye contact with a friend whom you pass on the street – in fact, different forms of eye contact signify different degrees of social familiarity. Fail to do any of these too often and one's social relationships will not last long.)

Sigmund Freud's (1856–1939) early **psychological** studies led him to liken public ritual to private obsessive compulsive disorders (OCD), such as repeated hand washing, and to compare collective **myths** to the psychological role dreams play in helping individuals to express otherwise repressed, anti-social desires in a public yet symbolic manner that does not threaten their place within the group or the future of the group itself. (Moral of the story: although it doesn't resolve the source of one's conflicted feelings for authority figures, it's easier to rebel against a symbolic authority than the actual one, since there are practical implications to rising up against the latter that don't attend rebelling against the former.) In light of his work, some today argue that religious ceremonies function to provide a safe venue for acting out and thereby satisfying anti-social urges, yet doing so in a tightly controlled, rule-governed, ritual setting so it does not get out of control and threaten the group, such as the ritual of sacrifice (whether merely symbolic, as in the Christian eucharist or communion, or actual), that we seem to find the world over – a ritual that may very well prevent violence from spilling over into all of social life by venting aggression on a symbolic victim who cannot retaliate (perhaps somewhat like transferring your aggression for a parent and, instead, 'taking it out' on a younger sibling?).

Current scholarship is pressing such classic work in entirely new directions, such as drawing on materialist scholarship and **semiotic** theory to study the political function of myth (e.g., **Bruce Lincoln**); using a social theory to account for such things as the beginnings of **Christianity** (e.g., **Burton Mack** and **William Arnal**); drawing on economic models of how people choose among alternatives (such as converting from one religion to another, as studied by **rational choice** social theorists such as **Rodney Stark**); and developing a theory of religion based on the findings of **cognitive psychology** (e.g., **Pascal Boyer**).

Of course, just as there are difficulties with the essentialist approach to definition (whether carried out from a theological or an anthropological perspective), so too there are problems with functionalist approaches – despite their current popularity. Among recent scholars of religion there may be no better example of a critic of functionalism than Hans Penner. Penner represents a group of contemporary scholars interested not so much in what religion, religious narratives or religious rituals *mean* but, instead, with *how things come to mean anything at all*. He is therefore not much interested in developing a theory of religion – as if religion was a *sui generis* thing that required a theory of its own either to interpret its unique meaning or explain its peculiar function; rather, he argues that scholars ought to turn their attention to developing a theory of meaning-making that can be applied to, among other things, those institutions we classify as religion.

Despite the fact that functionalism has now become the dominant approach in the human sciences, Penner nonetheless argues that it is deeply flawed, suffering from many of the same problems that plague essentialists. What has made functionalism so appealing to the last few generations of scholars was their apparent consensus that the source or the object of religious feeling defied explanation; because one could not get at the actual origin or the actual inner sentiments that were assumed to be the source of religious behavior, then – or so functionalists have argued – one has little choice but to shift one's focus and study the role played by the sentiments' expressions or what **phenomenologists** might call their manifestations (that is, the behaviors, the narratives, the symbols, etc.). Moreover, the role they play is assumed to meet some previously existing needs – whether those needs be sociological, psychological, political, economic, biological, etc. According to Penner, this shift from intuiting essences to observing functions, from non-empirical source to empirical manifestation, is no shift at all and the much celebrated gains of functionalism – such as the presumption that functionalists study something observable (that is, empirical), as opposed to the essentialist's non-empirical, subjective substance – are misleading or outright illusory.

That scholars have been unable to settle on which of a phenomenon's many functions constitutes *the* function is the first problem with functionalism. For there is no way to decide whether, to pick but two examples, a psychological or a sociological function is more basic, for the advocates of each seem to argue that their approaches are irreducible to those of their colleagues. Lacking a way of deciding, scholars seem to have settled for what Penner characterizes as a compromise position: the needs are now often assumed to be psycho-social as well as biological – hence, the interest, in the last generation or two, with representing the study of religion as a cross-disciplinary exercise that defied the limits of any one academic discipline. Because – or so it is argued – the expressions of the inner sentiment are many and the functions of these expressions are varied, a variety of tools are required for their study (from anthropology to literary criticism), none of which are sufficient on their own. Religion, it would seem, is a lot like love; to borrow the lyrics from the Oscar-winning theme song to the 1955 movie of the same title: it is a many-splendored thing. But, much as when, in the mid-1950s, China's Chairman Mao stated in a speech, 'Let a thousand flowers bloom, a hundred schools of thought contend', only to end up enforcing (by, in his case, the use of mass violence) a rather strict party line in the following years, so too despite the variety promised by the cross-disciplinary approach to the study of religion's functions, a rather narrow party line ends up being reproduced.

So, despite either their apparently **atheistic** or at least **agnostic** viewpoints, functionalists nonetheless seem to presuppose a basic **theological** belief held by the people whom scholars of religion study: that there is a non-empirical source to the behaviors and institutions that scholars study and that, because their methods require their object of study to be observable, scholars must set aside questions concerning this source, presuming it actually to exist but either to resist or confound their tools. Much as the phenomenologist, in bracketing the truth of the objects they study and, instead, describing only how the believer talks *about* their truth, ironically authorizes the believer's assumptions by leaving them untouched (and thus protected), so too the functionalist, in setting aside the sentiment and studying only the function played by its varied manifestations, further legitimizes the believer's assumptions by leaving them untouched (and thus protected). It is as if the functionalist grants the believer the authority to set the terms in which they are to be understood, merely complementing their claims by adding some observations on innate needs fulfilled by the believer's sentiments – observations that somehow escaped believers themselves. If this is the case, then what we can end up with is a field in which we study the adjective 'religious' but never the noun '**religion**'; that is to say, we are able to study its manifestations (e.g., religious symbols, religious ceremonies, religious sites, etc.) but when it comes to asking to what the possessive pronoun 'its' refers (i.e., 'its manifestations'), scholars must forever remain silent. We are therefore unable to answer a basic question that might have been posed to us by Plato's Socrates: what is it about religious beliefs, religious narratives, religious practices and religious institutions that they can all possess the same adjective? Simply put, what makes them religious?

An example of this difficulty with functionalism can be found in the work of Sigmund Freud; in the opening to his now famous 1907 essay, 'Obsessive Acts and Religious Practices' (published in his *Collected Papers* [1950]) (which makes the case for understanding ritual and obsessive actions to perform the same function for neurotics, though they may do it to different degrees) he had the following to say about definition:

> In place of a definition we must for the present be content with a detailed description of these conditions, for it has not yet been possible to demonstrate the essential feature which probably lies at the root of the obsessional neurosis, though one seems to find indications of it at every turn in clinical manifestations of the disorder.

On the one hand we seem to have this thing called 'the disorder' while, on the other, its various manifestations. Freud, then, seems to presume that there is some essential reality beneath its empirical indicators, much as a traditional literary critic assumes that behind, or beneath, the words on the page one can discover both the meaning of a text and the **intentions** of an author – both of which are obviously nonempirical and, because of this, thought somehow to remain constant despite the sorts of historial changes and accidents that we all routinely recognize to happen to printed texts, the paper on which print appears, etc. How, one might wonder, is such a position any different from an essentialist scholar such as, say, **Mircea Eliade**, arguing that, because 'the **sacred**' manifests itself in symbolic form, its study is not exhausted by examining its merely secondary symbolic manifestations?

It could be argued that functionalists try to **reduce** religion to non-religious causes, thereby failing to appreciate the true meaning of religious belief. Of course, it is not difficult to find this critique offered by those who wish to study religion either theologically or **humanistically**. Despite this criticism, for yet others, functionalists represent a traditional viewpoint, shared with essentialists, by studying material expressions and not their immaterial source, dressing their work up to look like something other than what it actually is.

5 The Public Discourse on Religion

Differences between the two approaches examined so far – the essentialist and the functionalist – make evident that, on the one hand, there are those who think that we can study religion by identifying private traits not readily available to our normal senses and, on the other, there are those who think that, like all other aspects of human behavior, the things we call religions have public features that perform observable functions. Given that we're trying to determine how best to study religion as an element of human culture, the distinction between the private traits and public actions deserves our attention.

When it comes to defining religion, there are thus two common approaches: one **inductively** browses through the group of things already known to be religions, looking for an **essentially** shared feature; and the other sets about looking for the common **function** performed in each religious practice or institution. If one takes the former route, then objects are defined by some key feature that is internal to them, more often than not some non-empirical feature judged to be *sui generis* (that is, self-caused, one of a kind, unique).

For instance, because there are innumerable observable differences among the members of the group known as, say, 'the British', people interested in defining what it is that they share in common – their 'Britishness', if you will – often fall back on the assumption that what really unites the members of this group is not some external feature (such as where they live, how they dress, what they eat, or how they speak – for these are all apt to change and the differences fly in the face of efforts to define the group's shared identity) but an internal experience, a feeling, an attitude, perhaps called 'the British experience'. This **experience** obviously cannot be tasted, touched, smelled or heard (that is, it is non-empirical), but, instead, only felt by the participants themselves and approximated rather crudely by the uninitiated observer or outsider (as in the response one sometimes hears in the midst of an argument: 'You don't know how I feel'). Because for many people religion is assumed to refer to an invisible but all too real interior world that is fully experienced only by the believer or the insider (a point often associated with **Otto**'s work), this essentialist approach is still very popular, within and outside of the academy. In fact, as the example of 'the British experience' makes plain, the popularity of the essentialist approach to defining social identity is not limited only to defining religion but also to defining national, gendered and ethnic identities as well.

But, if the category **'religion'** is to be used in the **human sciences** as a classification to name an aspect of the inter-subjective world – in other words, the public world that can be observed and studied, regardless of one's group membership or self-perception – then the essentialist approach is not very helpful for it is premised on the priority of a subjective, private world of affectation and aesthetic appreciation that is presumed to be unavailable to non-participants. Because the functionalist approach focuses on the use to which something is put, it shifts attention to defining something in light of an observable group of people, their needs, their actions, and their practical interests. (Seeing these as all interconnected, thereby requiring nuanced fieldwork to observe the system in its entirety, was a basic assumption of what, in **anthropology**, was once called the social functionalist approach.) A number of scholars therefore argue that the functionalist approach holds more promise for the academic study of religion practiced as part of a public discourse. That some scholars are still troubled by the internal, non-observable nature of functionalists who talk about 'needs' as if they were internal things that inhabited all human beings, should not be forgotten. But at least the switch to a functionalist approach attempts to remake the study of religion into something that can take place as part of the public sphere.

At least when thinking of the US, where I live and teach, the topic of religion and the public sphere immediately brings to mind the place of religion in politics and the courts. Although the study of religion, like many of the other disciplines in the modern university (such as **anthropology, sociology, psychology**, etc.) first arose in Europe in the late nineteenth-century and came to North American universities prior to World War I and, for a brief time, flourished at such schools as the University of Chicago, the University of Pennsylvania, and Harvard, it was not until the late 1950s and early 1960s that departments of Religious Studies were established in most US public universities – institutions that, because they are tax-supported, are required to meet certain obligations specified in Federal law, such as the so-called 'separation of Church and State'. The establishment and success of these departments in the public university system can be related to a number of factors: the increasing interest in Asian cultures brought on by such events as the Korean and then Vietnam wars (or were they 'police actions' and not wars? My point? Classification matters!); the way religion (and specifically, certain forms of **Christianity**) was used in US political discourse during the height of the Cold War as a way to distinguish some posited 'us' from those 'godless Communists'; the country-wide changes taking place in the Civil Rights era, in which the balance of traditional power was slowly beginning to change; and, given changes in US immigration policies in the mid-1960s, the growing presence of new minority populations that came from places other than Europe (which previously comprised the US's main source of immigration), bringing with them foods, clothes, names, and practices that must have struck many of those already settled into life in the US as, well, exotic (maybe even dangerous and thus threatening?). It therefore seems reasonable to expect that, in such a context, university classes would, sooner or later, begin to focus increasing attention on describing other cultures and looking for cross-cultural similarities amidst the obvious differences. After all, the very origins of the field almost a century before can easily be linked to the same sort of rise in speculations about 'other people' among the curious, even perplexed and possibly shocked, Europeans reading reports of travelers, missionaries, and military officers who were part of their countries' **colonial** missions. For it's rather tough to maintain your own sense of having an utterly distinct and thus superior culture when you realize that certain things you had taken for granted as being unique to yourself appear to be practiced, with varying degrees of difference, worldwide. This presented a problem, of course, to people who had seen those outside Europe as being nothing but uncivilized savages who had not known the grace of God. Questions began to be asked: is our culture the same as all the others? Being a tough proposition to entertain for our intellectual predecessors, it became appealing simply to concede just a little and to assume that all human cultures developed – or better, evolved – but at differing rates, with one's own being nowhere but at the head of the line. This we saw in **Tylor**'s theory of **animism**, drawing as it did on **evolutionary** assumptions.

Jumping ahead to the mid-twentieth century, the effort to establish the field in the US public university was given momentum by the US court system; for, given the system of government in the US, sooner or later public disputes tend to end up in the court system. And so, with increasing difference within the population – or at least new differences previously unknown by the majority of the population in the mid-twentieth century – we find courts in the late 1950s and early 1960s addressing a variety of **Church/State** issues.

To discuss the role of religion in US public discourse one must therefore understand the Supreme Court's reading of the US Constitution, especially the Bill of Rights (the name given to the first ten amendments, adopted by Congress on December 15, 1791). The famous opening lines to the **First Amendment** to the Constitution read:

'Congress shall make no law respecting an establishment of religion, or prohibiting the free exercise thereof...'

Legal scholars distinguish between the First Amendment's **establishment clause** and its **free exercise clause**: the Amendment is interpreted to state that the elected government cannot enforce, support, or encourage (that is, 'establish') a particular religion (or religion in general, some would argue), nor does it have the right to curtail its citizens' religious choices and practices (that is, the 'free exercise' of their religion). (Before proceeding, it is worth considering, for a moment, what emphasis to place on the word 'respecting' in the establishment clause – does it mean 'with regard to', in which case its presence here is of little consequence, or does it mean 'to have regard for', as in 'to honor', in which case it prevents government not simply from establishing but even seeming to favor religion or appearing to sanction the establishment of a state-supported religion. As you can see, much therefore depends on how this one word is interpreted.) Of interest to some is that, in the opening lines of the First Amendment, it is made explicit that all citizens of the US have the absolute right to believe in any or no religion whatsoever. This was a point upheld in an April 28, 2005, press conference by US President, George W. Bush, who, although well known for often speaking publicly about the importance of his own Evangelical Christian 'faith', said in reply to a reporter's question concerning how religion was being used in US political discourse: 'The great thing about America is that you should be allowed to worship any way you want, and if you choose not to worship, you're equally as patriotic as somebody who does worship.'

Although this once taken-for-granted, or at least dominant, view on the separation of Church/State is now being reconsidered by some within in the US – as evidenced not only by such recent Federal programs as President George W. Bush's **Faith-Based Initiative**, but by the very fact that the President apparently saw it necessary to specify in his press conference that one's religious faith was not necessarily a measure of one's patriotism – his comments presuppose elements found in the majority decision in a landmark 1963 US Supreme Court case. This case – one among several from this era that have had an enduring impact on life in the US – concerned the School District of Abington Township, in the state of Pennsylvania, which was being sued by the Schempp family whose children attended one of its schools; the outcome of this case, which was eventually appealed all the way to the Supreme Court, helped to change the way public schools operate, and its effects are still felt to this day, most notably when it comes to how delicately religion is treated in US elementary and high schools. In this case, a family successfully sued a public school board for one of its school's daily opening exercises – which, at that time in US history, would routinely include such elements as reciting the **Christian** 'Lord's Prayer' and readings from the Bible over a school's public address system at the start of each school day. The Court decided that, as a publicly funded institution charged to represent and not exclude the members of a diverse, tax paying citizenry, the school board was infringing on the rights of its students, not just by supporting a specific denominational **worldview** but, more importantly perhaps, a religious worldview in general. Therefore, the Constitution's establishment clause was the topic of concern to the Court.

Justice Clark, the Supreme Court Justice who wrote the decision in this case, on behalf of the majority of judges, stated that although confessional instruction and religious indoctrination in publicly funded schools were both unconstitutional (because they infringed on the First Amendment's restrictions on government 'respecting the establishment of religion'), one's education, Justice Clark famously went on to write (in a line often quoted by scholars of religion at that time), 'is not complete without a study of comparative religion or the history of religion and its relationship to the advancement of civilization'. The majority of the Justices therefore interpreted the First Amendment to state that, although the government cannot force a student to be either religious or nonreligious, the government certainly can – and, or so Justice Clark seemed to argue, probably should – support classes that study the history of particular religions, the comparison of two or more religions, and the role of religion in human history.

Basic to its decision was the Court's distinction between religious instruction and instruction *about* religion – a distinction popular among scholars of religion writing in the 1960s and 1970s (notably such writers as **Ninian Smart**) who had the task of persuading their colleagues, who were already studying religion in such long-established academic disciplines as **Anthropology**, **History**, **Sociology**, etc., that there also ought to be a place in the modern university for the academic study of religion as a distinct field (what, in North America, quickly came to be known as Departments of Religious Studies – though 'Department of Religion' and 'Department for the Study of Religion' are also names used by some schools). The academic study of religion as practiced in a public context is, they argued, concerned to study *about* religion and religions and not be a form of religious practice (whether theologically liberal or conservative). Regardless of how it is practiced in the many private institutions found throughout the US – whether that means elite or so-called Ivy League universities or small denominationally affiliated, liberal arts colleges – it was argued that the study of religion in public institutions ought to follow the same rules of argumentation, rules for the use of evidence, and ways of gathering data (what we might call its 'methods') as other intellectual pursuits. And, in many institutions, this argument won out.

So, we find ourselves today with three different ways of studying religion: **theological** studies of various sorts, housed in private and denominationally supported schools, which is aimed at articulating, in a systematic and rational manner, the principles of the participant's viewpoint (whether that viewpoint is mainline or marginal, whether theologically liberal or conservative); **humanistic** studies that, after comparative work uncovers deep similarities, understand diverse participant viewpoints to share universal values that are not necessarily apparent even to the participants; and those in the **human sciences** who understand all claims concerning the existence of such things as deep essences, self-evident meanings, and universal values to comprise an instance of data – whether that be claims concerning the existence of souls or the Human Spirit. But despite these very real differences, one thing seems to be shared by each approach: they all presume the difference between participants and non-participants, whether that be those who are saved and those who are not, or those who practice religion and those who study it. In two words, they all presume the difference between *insiders* and *outsiders*.

6 Religion and the Insider/ Outsider Problem

It is clear that there are implications for our studies if we presume religion to be an inner trait, sentiment, belief, or experience that can only be expressed publicly in some secondary manner. For in this case, the actual nature of religion always eludes the observer's grasp – since they are left with what some have called its secondary aspects or its externals – making the academic study of religion an impossibility. Implicit here is the distinction between how participants understand this thing we call religion and how non-participants understand it, suggesting that in order to study religion we need to develop some tools to distinguish these two viewpoints from each other.

The commonly described distinction between studying *about* religion and studies that are religious (or **theological**) brings to mind what is commonly called the insider/outsider problem – an issue present in Andrew Scott Waugh's efforts to use only local names for the mountains he identified during his early mapping of India (although, in naming Peak 15 after Colonel Everest, he hardly followed his own rule). Because much of the original work on the insider/outsider problem was done in fields outside the academic study of religion, it is only fitting to open a discussion of the insider/outsider problem with reference to the work carried out in the field of **Linguistics**.

From linguists, anthropologists and then scholars of religion borrowed two technical terms – **emic** and **etic**. These two terms roughly correspond to experience-near and experience-distant, terms used by the US anthroplogist, **Clifford Geertz**, to suggest the continuum that may exist between those **experiences** that are familiar and those with which one has trouble identifying. Like experience-near and experience-distant, emic and etic are concepts that are commonly used when tackling the study of meaning systems outside of one's own, whether they are language systems, cultures, economies or religions. Although both words are derived from the same **Greek** root (*phonema*, meaning sound, hence our words phonograph and telephone), the term *phonemic* designates the sounds themselves (or what are called phonemes, the smallest units of meaningful sound in any language system) whereas the term *phonetic* specifies the symbols and the organizational systems that scholars devise to represent and then compare the manner in which the basic phonemic units of language systems are pronounced and strung together in complex relationships that are commonly known as words or sentences. To the proficient users of any language (the linguistic insiders who possess the emic perspective), studying these systems (the etic perspective) may or may not be an interesting topic; after all, they are involved in using, articulating and developing a language for certain practical purposes and likely have little interest in developing a **theory** of human language as a general phenomenon or a systematic study of how their discrete units of sounds are produced and how this compares to the production of sounds in other languages. To these language users (also termed native informants, participants, or actors) the varied ways of producing a unit of sound that they understand to be significant (or, say, producing any symbol or action that they understand as meaningful and worth remembering) might all just appear to be self-evident, ordinary, and thus uninteresting – after all, readers likely don't think twice about what their diaphragm, lips, tongue and teeth are all doing when they make the sound represented by the symbol 'T'. But to a novice, such as a non-Spanish speaker trying to master the trilled (or rolling) 'R' sound that is used in Spanish but not in many other languages, or to the theorist of human-language-in-general or human-**culture**-in-general, the ways in which subtle distinctions in performance and meaning are produced by speakers or social actors, and the ways in which these subtle distinctions are understood to be significant by those who receive them, can be intriguing.

Before continuing, an important point must be made: it seems likely that insiders do not actually have a viewpoint, as much as they simply go about their business, fully immersed in their particular meaning/behavior world (just as readers of this very page are fully immersed in the rules of English grammar, or much as with the implicit assumptions of people replacing light bulbs and starting car engines). Therefore, the emic or insider perspective might better be understood as the outsider's attempt to reproduce as faithfully as possible – in a word, to *describe* – what might be considered to be the informant's own descriptions of his or her production of sounds, behavior, beliefs, meanings, institutions, etc., should they offer such an account. I say 'what might be considered' because in many fields there is no insider's view against which to compare our descriptions, to see if we got them right, for many of the people we study are long gone and all that remains are difficult to interpret artifacts recovered by archeologists. (Question: Is this artifact a toy or an idol? What will people long in our future make of what they find left over from our culture? Will a Barbie doll, which will surely not degrade very quickly, be understood as an idol? Once again, classification matters!) But more than this: even when the insider is present, such as a native informant being studied by an **anthropologist** doing fieldwork, often they go about their daily lives in such a way that they are not consciously considering the motivation, meaning and implications of their actions – all of which no doubt interests the observer who busily asks them questions about these things, ensuring that the answers provided by the participant (once they are translated, that is) are in fact a product of the non-participant's curiosities. For instance, as just indicated above, although readers of these words are likely more than capable of answering questions about the significance of their ability to decipher the scribbles on this page by means of their knowledge of the English language's alphabet and its grammar, yet in the midst of this decoding they likely are not much thinking about their language abilities or the implications of being literate. Instead, they are unreflectively immersed in their world. If this is the case, then the etic perspective is the observer's subsequent attempt to take their own descriptive information (emic) and to organize, systematize and compare – in a word, to *redescribe* – it in terms of a system of the scholar's own making (such as the cross-culturally useful phonetic alphabet, which as-

signs local symbols to sounds made cross-culturally or the scholarly definition of the technical term myth that is used to distinguish and sort the
many types of narratives people tell and perform). What should be clear,
then, is that the emic and the etic perspective are deeply intertwined.

So, despite what some might think, the emic perspective does not
exist apart from the non-participant's curiosity (to rephrase: without outsiders I'm not sure there would even be an insider viewpoint); insider
disclosures about such things as the causes, meaning and implications of
their social world are therefore solicited by the queries of the curious
non-participant who is familiar with a different world and therefore curious about this or that new sound, belief, action and institution, etc. (This
is a point basic to much of **Jonathan Z. Smith**'s work: gains in knowledge
come about by means of the curiosity that results when we inevitably
learn that the world does not always work as we think [might we say
guess? Or perhaps hypothesize?] it does.) It is therefore crucial not to
confuse the emic perspective with the so-called insider's own actual viewpoint, as if our descriptive work somehow allows us to leave our own
bodies and read other people's minds (mind reading attempts can sometimes be quite common among scholars studying other people, somewhat like Tylor thought he could conjure up the 'savage philosopher's'
thoughts); for, as in the case of language, language users are extremely
proficient at speaking their language, at making this or that sound (i.e.,
phoneme) in a manner that is distinct from other sounds (for example, it
is crucial, in English, to produce the proper phoneme so as to distinguish
the vowels in such words as 'fit' and 'fat', especially when trying to com-
pliment someone by saying, 'You look particularly fit today'), but such
people are often hardly interested in first reflecting upon how it is that
they do it and then comparing their techniques to those of other language
users so as to answer questions about the feature of language in general
as a human phenomenon. By even attempting to *re-produce*, rather than
simply *produce*, a sound faithfully, the linguist has already acknowledged
that she or he is a student of the language under study (could we call this
studies *about* language?) and is not to be confused with merely being a
speaker of the language or even someone trying to learn to speak the
language (that is, someone trying to learn to master the art and thereby
reproduce the various phonemes).

It would seem, then, that the insider/outsider problem – phrased simply, the question of whether, and if so, then to what extent, non-participants can study others – involves a complex continuum of positions and
viewpoints. If we add to this the problem of who gets to count as an
insider – such as English language users who distinguish between 'The
Queen's English' or so-called 'proper English', on the one hand, and other

so-called degraded forms of English (as in the vernacular, or common speech) – it is clear that *there are insiders and then there are insiders!* So who gets to be *the* insider? As anthropologists who do fieldwork found out a generation or two ago, one cannot simply presume that talking to the senior males of the village exhausts all of the information to be gained from studying a group, for there are many sub-groups and it is highly likely that not all share the same perspective and interests. Case in point: for anyone keeping abreast of the news since the attacks of September 11, 2001 – in which politicians and commentators worldwide, on all political sides, have debated which form of **Islam** is the 'true faith' and which forms are 'deviant' and 'fanatical' – it should be obvious that the stakes in the insider/outsider debate can be quite high and are all too real.

Therefore, instead of simply assuming that there is a stable and self-evident insider and a clearly distinguishable outsider viewpoint, or instead of trying to judge which of a group's many forms is its supposedly the authentic heart, the pristine original, and its authoritative form, the questions that some scholars have now begun to ask are: Which of the many viewpoints is to be authorized? Based on whose criteria? How does this authorization take place? Is the quality or accuracy of etic scholarship to be judged by the informant's criteria? (In other words, is 'Mount Everest' illegitimate because it is not a local name?) Do people doing some behavior or holding some belief set the terms by means of which their behaviors and beliefs are to be understood? Or is the informant (as an anthropologist might name the insider) to be judged by the conclusions reached by the observer? Does scholarship operate apart from the concerns and interests of insiders or is it intimately connected to their lives? What if the insider disagrees with the scholar's conclusions – as has begun to happen worldwide now that the writings of scholars are easily available to many of the people on whom they write? Is the goal of scholarship on human behavior, beliefs and institutions to have the people whom we are studying agree with our conclusions and generalizations, or is it instead the goal of gaining new knowledge by developing logical, scientific theories concerning why it is that humans do this or that in the first place, regardless of what people report their own motives to be? Simply put, to whom do scholars of human behavior answer and what is the purpose of their work? All this is entailed in current discussions of the insider/outsider problem.

In arguing for a theory of religion in general, especially when carrying out this type of research from within a public university (which is funded through the tax dollars of an obviously diverse citizenry, comprised of all sorts of **theistic** viewpoints, as well as those of **agnostics** and **atheists** alike), it was maintained that no one religious viewpoint could come to dominate, for we are not attempting to develop a **Christian** theory of religion, a **Jewish** theory of religion, a **Hindu** theory of religion, a **Muslim** theory of religion, or a **Buddhist** theory of religion – though no doubt theologians in each of these movements will likely wish to develop their own views on this topic. Instead, much like those who study phonetics, scholars of religion seek to develop criteria from outside each of these particular social systems so as to compare them or their elements, and then to develop a theory capable of explaining the possible reasons for the similarities and differences that their comparisons bring to light. Etic scholarship therefore ought not to be constrained by the way in which the people one studies say they act or think. Instead, it should be constrained by the rules that govern all rational, comparative, scientific analyses that are found throughout the modern university.

To explore the insider/outsider problem a little more, consider the following example, which might arise in the study of those mass social movements that we commonly call religions: the problem of evil. Commonly called **theodicy**, this is the theological challenge of justifying one's belief in an all-powerful, all-good god (also called a deity) despite events in the observable, natural world (everything from disease and crime to natural disasters) that suggest otherwise. This has been a long-standing problem for some theologians who subscribe to belief in a monotheistic god (the belief that there is only one god). Therefore, any scholar of religion interested in studying the history of a particular type of theology, or current theological debates, might wish to survey, among the many other topics, what people are saying about the problem of evil. But in carrying out such a study, are scholars of religion studying evil, studying the problem of evil, or studying, first of all, specific groups of people who believe evil exists and, second of all, their attempts to deal with the apparent conflicts between this and other of their beliefs? Simply put, is the category 'theodicy' emic or etic? To rephrase, for whom is the problem of evil a problem? The scholar of religion or the theologian?

The scholarly model we might wish to draw upon in answering this question is that of **Emile Durkheim**. For Durkheim, participants' claims about such things as gods or the origins of the universe were understood as coded statements about the social order itself. Adopting a **naturalistic** approach, he understood claims about God to be a way in which participants projected outside themselves their sense of the group's own needs, making religion – or, more specifically, **mythic** narratives and public **ritual** performances – but two among the many sites at which society, which exists nowhere but in the minds of its members, continually renews itself. Therefore, what for insiders counts as talk about God, God's wishes, and God's judgment is, for the sociologist, symbolically coded talk about society, society's wishes, and society's judgments. Although contemporary social theorists no longer employ Durkheim's understanding of society as being a monolithic whole, and instead assume that the thing once simply called 'society' is, in fact, comprised of a host of subgroups, with overlapping memberships, that conflict and complement each other in a variety of ways, they would more than likely agree with Durkheim: upon analysis, religious discourse turns out to be the language of social life. Religion is **reduced** to society.

But what has this got to do with theodicy?

In order for the problem of evil even to be *a problem*, to be something worth puzzling over – that is, to be an item of **discourse** – one needs a few assumptions up and running, assumptions that are part of the participant's world but not necessarily part of the non-participant scholar's world. In fact, not all people identified as religious participants might even have the assumptions necessary for the problem of evil to appear to be a problem worth discussing. Christians, of course, have certainly been preoccupied with this topic, but not necessarily everyone designated as religious develops theodices. So, right off the bat, we need to entertain that only some of the people we study develop rational systems we might call theodicies – suggesting that, because the term 'theodicy' may have limited cross-cultural use, it is hardly an etic category that will be of use in our work as scholars of religion.

But for those who do develop them, theodicies presuppose a belief that such a thing as evil exists and, second, it must be understood to be in conflict with some other way in which things could be working out. If, instead, one presumed that the universe was itself a random outcome of countless unintentional events (and notice I did not say 'actions', so as not to imply some **agent** doing something for a reason and an intended outcome), then it might not make much sense to talk about evil. Sure, given one's own set of interests (such as wanting to live a long life) one might still deeply lament this or that outcome but the outcome's conflict with one's own hopes and dreams would hardly constitute an instance of what we commonly designate as evil. Things could still be judged unexpected, shocking, even horrific, but such value judgments would always be understood to operate in relation to a specific observer's point of view, making shock an item in the eye of the beholder (that is, it is a function of his or her expectations) rather than the name given to some **essential**, inner value the thing has always possessed. In this case, there is no evil, for things could not have been otherwise – in other words, conflicts, gaps and disappointments, some great and some trivial, are part of the very fabric of historical existence (as an **Existentialist** might say). Therefore, although deeply frustrated people may remain, there is no problem of evil.

To study the problem of evil, to try to solve the problem posed by the existence of evil, therefore means one assumes that things ought to be otherwise, that things ought to make sense, that things ought to work out according to a certain sort of pattern. The problem, then, is that they often (generally?) do not. Failing to assume that things ought to make sense, one can develop ways to study other people's assumptions about how things ought to work, along with studying their attempts to explain the discrepancies between their hopes and dreams, on the one hand, and the ways things apparently turn out at the end of the day, on the other.

This means that theodicy is an item of a particular sort of emic discourse – often, an economically and intellectually elite emic discourse, for it is likely that most people, when confronted with a shocking and unexpected turn of events, do not bring to bear an elaborate, rational system to account for the event in light of a belief in a powerful, good God. Instead, they likely bear up under the circumstances as best they can, concerned in the short term with profoundly practical concerns. Later, if they happen to have the luxury of acquiring some sort of physical or intellectual distance from the event (making them an outsider to their own, earlier status as an insider to the experience), they might reflect on such things as the meaning of the event – something they determine by placing it in relation to assumptions they routinely employ while going about their daily life.

Of course, non-participant scholars who adopt the cross-cultural, etic viewpoint, and who happen to be interested in this topic, will surely wish to describe all of this, in great detail, and will likely wish to compare different attempts to grapple with the issue, and in the process examine how participants distinguish between better and worse attempts to address the perceived problem. But scholars will more than likely not be content with accomplishing just this descriptive work. Instead, they might eventually wish to account for why people think the universe ought to work in this or that fashion, and why they opt for belief in a powerful being as a way to answer their questions concerning why things don't go in their favor more often. Perhaps, like Durkheim, they'll employ a social theory in their explanation, concluding that theodicies are actually coded social language, a technique for justifying not the ways of God but the ways of the social group. Or perhaps they'll draw upon work carried out in a field such as **cognitive sciences**, seeing theodicies instead as ways in which social actors articulate and systematize their efforts to deal with what we once might have termed cognitive dissonance – a term used for the discomfort, and sometimes profound anxiety, that results from having experiences that conflict with one's conceptual framework.

Regardless which etic framework one adopts to study not the problem of evil, but the discourse on evil, it should be clear that, although they might overlap greatly on the descriptive level, participant and non-participant discourses vary greatly after the work of description has ended and explanation and analysis have begun.

7 The Resemblance among Religions

With private and public, and insiders and outsiders, distinguished, we can return to the problem of definition, to consider a third approach that many scholars think avoids some of the shortcomings of the essentialist and the functionalist approaches. This approach is commonly found in textbooks today, and is known as the family resemblance approach to definition.

Readers might recall that, in the discussion on **essentialism**, they were told that a light could either be on or off, and never partially on or partially off. So too with essentialist definitions. In fact, the presumption that there is a distinct insider perspective as opposed to an outsider view – as opposed to seeing insides and outsides as continually changing and continually contested, all depending on where you stand and in relation to whom – is itself a product of an essentialist viewpoint. Although the light switch imagery works to communicate the 'either/or' nature of this approach to definition (whether defining who gets to count as a 'patriotic citizen' or who counts among 'the faithful'), surely some readers must have thought, 'What about that dimmer switch in the dining room?' Good point. In the study of definitions the dimmer switch example would be called the **family resemblance** approach.

The family resemblance approach to definition – sometimes called **polythetic** definitions – is thought by some to enable them to steer a middle path between essentialist and **functionalist** approaches. Although one of the most useful contemporary examples of this approach would be the 'dimension theory of religion' advanced by the **phenomenologist** of religion, **Ninian Smart**, this approach is credited to the philosopher, **Ludwig Wittgenstein** (1889–1951), who, in trying to argue that there is no one defining characteristic that makes something 'language', asked his readers to stop and consider how it is that they actually go about the activities of classifying, sorting and distinguishing things as specific types of things. If they did this, he suggested, if they actually considered what it was that they did when they went about the activity of naming and demarcating, they would see that all members of a particular group *more or less* share a series of traits or characteristics. To put it another way, all members of a common group overlap to varying degrees and in differing respects, just as no two members of a family are exactly alike (even identical twins differ); instead, they more or less share a delimited series of characteristics (such as name, hair color, temperament, height, favorite foods, blood type, etc.). Further, despite some people's best efforts to portray themselves as authoritative, no family member constitutes the definitive instance of the group – rather, all members share in the identity, to varying degrees. Group membership, Wittgenstein argued, is never a matter of yes or no (as in the essentialist approach) but always a matter of degree, a matter of 'more or less'. Anyone who has had a person marry into their family, or who has married into another family, knows as much: it takes time to become a member of a group. Only once a series of customs has gradually been learned and mastered (Where do I sit at the dinner table? Which way do I pass the potatoes? Do we say grace?

When do I start eating? Can I ask for more milk?), can one's membership be taken for granted.

To make his point, Wittgenstein uses the example of games, asking his readers to consider the variety of activities we commonly know by this name, challenging them to come up with the one essential trait that all of these activities share. '[T]he result of this examination', he concludes, 'is: we see a complicated network of similarities overlapping and criss-crossing... I can think of no better expression to characterize these similarities than "family resemblances"; for the various resemblances be-tween members of a family: build, features, color of eyes, gait, tempera-ment, etc., etc., overlap and criss-cross in the same way. And I shall say: "games" form a family'. The acts of classification and definition, for Wittgenstein, were therefore activities of selection, of choice, not of merely passively recognizing qualities thought to reside in some object that catches our attention (such as the essence of beauty, of justice, or of religion); instead, something was only 'more or less' this or that (hence the usefulness of the image of the dimmer switch on the ceiling light). So, if only two traits are shared in common between two objects, say those two things we call an apple and an orange (they are both round and can be eaten), we have no choice but to make a judgement call regarding whether these two things are the same. But are only two traits sufficient? If not, then what about three? Or four? Case in point: is **Buddhism** a religion, despite the lack of any belief in a god in some of its forms? Or what do we make of some Evangelical Protestants who assert that Roman Catholics are not **Christians**? They seem to be saying that their Roman Catholic counterparts are not part of the family that they themselves know as 'Christian', likely due to the latter group being judged by the former as not sharing a sufficient number of traits.

If we follow Wittgenstein, then it seems to fall to those who develop and use classification systems – such as those who attempt to define religion – not only to have what a recent **anthropologist**, Benson Saler, has termed a 'prototypical definition', but also to be prepared to make judgment calls when a cultural artifact meets so few of their prototype's characteristics that it is questionable whether the artifact can productively be called a religion. That the **prototype** we use when we set about defin-ing religion in this manner is often confused with being the ideal case or the norm is certainly a trouble of which scholars ought to be aware if they wish to avoid making but one social movement the norm. It is just such a confusion between prototype and ideal that has sometimes led European and North American scholars to use certain types of Christianity or **Islam** as the authoritative standard by which they measure the quality and legiti-macy of other social movements also known as Christian or Muslim.

Contrary to essentialist and the functionalist scholars passively recognizing either some core feature or purpose/need served by a religion, Wittgensteinian scholars of religion see themselves as actively constituting a cultural practice *as* religious insomuch as it does or does not match the prototype which they understand as their starting point. Moreover, they are prepared to adjust their prototype for it is merely a tool and a starting place. That the family resemblance definition widens in the case of more liberal scholars (either politically or theologically), and narrows in the case of those who are more conservative, should not go unnoticed.

So what might a family resemblance definition of religion look like and do such definitions also have shortcomings? Consider the following two examples, the first from the philosopher, William Alston, and the second from the **historian of religions, Bruce Lincoln.** Alston, in the entry on 'Religion' in the *Encyclopedia of Philosophy* (1967), defines religion by means of what he characterizes as 'religion-making characteristics':

1. Belief in supernatural beings (gods).
2. A distinction between **sacred** and **profane** objects.
3. **Ritual** acts focused on sacred objects.
4. A moral code believed to be sanctioned by the gods.
5. Characteristically religious feelings (awe, sense of mystery, sense of guilt, adoration), which tend to be aroused in the presence of sacred objects and during the practice of ritual, and which are connected in idea with the gods.
6. Prayer and other forms of communication with gods.
7. A **worldview** or a general picture of the world as a whole and the place of the individual therein. This picture contains some specification of an overall purpose or point of the world and an indication of how the individual fits into it.
8. A more or less total organization of one's life based on the worldview.
9. A social group bound together by the above.

Bruce Lincoln, in his book *Holy Terrors: Thinking about Religion after September 11* (2003), defines religion as follows: 'A proper definition must therefore be polythetic and flexible, allowing for wide variations and attending, at a minimum, to these four domains':

1. A **discourse** whose concerns transcend the human, temporal, and **contingent**, and thus claims for itself a similarly transcendent status...
2. A set of practices whose goal is to produce a proper world and/or proper human subjects, as defined by a religious discourse to which these practices are connected...
3. A community whose members construct their identity with reference to a religious discourse and its attendant practices...
4. An institution that regulates religious discourse, practice, and community, reproducing them over time and modifying them as necessary, while asserting their eternal validity and transcendent value.

Although hardly the same – Alston emphasizes rather traditional aspects associated with religions (e.g., an emphasis on belief) whereas Lincoln focuses on religion's political role in establishing authority and order – both definitions offer a series of characteristics, or domains, that one would expect to find when looking for a religion. It is held that the advantage of this way of defining is that it is thought to avoid **reducing** religion to some essential trait or function; for, as Lincoln – citing the work of **Talal Asad** – warns, '[a]ny definition that privileges one aspect, dimension, or component of the religious necessarily fails, for in so doing it normalizes some specific traditions (or tendencies therein), while simultaneously dismissing or stigmatizing others'.

But, just as with the essentialist and the functionalist approaches, a few criticisms are possible here as well. First, family resemblance approaches often appear to be circular definitions (what we might call tautological or repetitive); religion, for Alston, is defined in light of such words as 'sacred' and 'profane' – words that are themselves generally defined in light of religion, suggesting that we already need to know something about religion in order to understand the very words used to define religion – hardly a helpful approach to defining something. For instance, Alston's fifth point defines religions as having 'characteristically religious feelings', and Lincoln's definition defines religions as being comprised of 'religious discourses' – defining the noun 'religion' by means of qualifiers that employ the adjective 'religious' likely does not help us too much.

A second, related difficulty involves the role played by one's prototype; when one reads a family resemblance definition, it often seems to be no more than a description of the thing one is defining – for example, it is not difficult to see Christianity lurking behind Alston's definition. It is as if the scholar was looking at a religion, describing its main features, and then generalizing from these to establish a definition of what one would expect other religions to possess. As Benson Saler has made clear (in a book on defining religion entitled *Conceptualizing Religion* [1993]), we have no choice but to employ prototypes – after all, we never experience the concept of 'table', but, rather, we seem to experience actual tables, one or more of which we seem to use as the model for identifying those other things we classify as tables. But, as I can imagine Socrates asking, precisely how did we know that the thing we had first encountered – that which became our prototype – was a religion? Did we recognize in it some deep, essential feature? Or were we simply told that it was a religion, thereby simply adopting this classification since childhood? How, for instance, did Alston know that Christianity was a religion and that its features could also be found elsewhere? And what if something else had

constituted his prototype? What then would his definition look like and
how useful would it be to us in carrying out cross-cultural research?

This is the criticism developed in an article published in the British/US
academic journal *Religion* in 1996 by the British scholar of religion, Tim
Fitzgerald: despite the fact that the family resemblance approach is por-
trayed as more inclusive and therefore capable of recognizing the variety
of actual religions, as with Penner's critique of functionalism, there is an
odd sense in which an essentialist definition yet remains at the very core
of the family of traits by which other things are identified *as* religions. For
Alston's definition seems to be saying: Christianity is a religion; Christianity has these various components; therefore anything else that is a religion will also have some or all of these components. In the midst of all
this, the essentially religious identity of the thing we call Christianity is
simply assumed and thereby reinforced. Why it gets to count as a religion
– rather than, say, a mass socio-political movement – is never explored.

The problem is that, as already identified with the insider/outsider
problem, such an approach merely takes what one group of participants
already understands as a religion (what we might call their prototypically
folk knowledge) and extends that (often in a rather vague manner) to
other cases, as if generalization was all that was required to turn folk
knowledge into a scholarly theory. In so doing, the scholar implicitly authorizes one among many (potentially competing) **emic** perspectives.
After all, people the world over use a host of local, emic concepts to
name and thereby distinguish the parts of their social worlds that strike
them as important, yet scholars would never think of using just any local
concepts as if they all named cross-cultural, human universals. So why are
we so willing to do this in the case of our familiar word 'religion', especially if we, as scholars, don't do any retooling and, instead, simply use it
in its common folk sense?

8 Religion and Classification

It should now be apparent that all three common approaches to definition – essentialist, functionalist, and family resemblance – have problems associated with them. But don't despair; for if definition is a human action used to make the world knowable and thinkable, then one wouldn't be surprised to find that the very tools we use to define things have limitations of their own. After all, even multi-purpose Swiss army knives, complete with can openers and corkscrews, can't be used for everything. Why? Because despite doing their best to anticipate eventual needs, their designers are not all-knowing and their tools inevitably fall short because the interests of the tool's users continually evade the designer's knowledge.

What's the point? If tools are devised to accomplish interests, but if those interests are forever in motion, then the devices that we use to make the world knowable (those things that we commonly call concepts, categories, systems of classification, etc.) must continually be re-tooled – and sometimes even discarded – all depending on our ever-changing needs. So the category 'religion' is your tool; how will you define it and what are you going to do with it?

Keeping in mind the relationship – suggested from the outset by **Mary Douglas**'s comments on 'soil' and 'dirt' – between classifiers, their system of classification, and that which they are classifying, we can see why a number of contemporary scholars have found the **essentialist** approach to be unproductive inasmuch as it presumes a common identity, or essence, to underlie a thing's many varied manifestations – the presumption that motivated an earlier scholarly movement known as the **Phenomenology of Religion** (e.g., see the Dutch scholar, **Gerardus van der Leeuw's** classic 1933 work, *Religion in Essence and Manifestation*). Classification is now seen by some to be an inherently and inescapably political activity – something apparent to some members of the United Nations when, not long after the attacks of September 11, 2001, and during debates on developing a means to define and then combat terrorism, they made a point of arguing that groups considered 'terrorists' by one nation might just as easily be conceived as 'freedom fighters' by another – all depending on how closely their goals do or do not complement those of the classifier. (This might explain why there is currently no agreed-upon definition of terrorism among the UN's member states.) So, just as studies of the politics of scholarship have recently appeared throughout the **human sciences**, so too the study of religion is being re-conceived as a site constituted by choice and practical interests rather than one based on sympathetic spiritual insight (see, for instance, the work of **William Arnal, Talal Asad, Bruce Lincoln** and **Tomoko Masuzawa**).

But as we have seen, **functionalist** and **family resemblance** approaches to defining religion can also be judged to be insufficient. Does that mean that, as so many scholars before us have concluded, religion cannot be defined – that, as someone like **Rudolf Otto** might have concluded, due to its complexity and subjectivity, it can only **experienced** and insufficiently expressed? If so, then how can one *study* it rather than just *feel* it? As many scholars have asked, if religion is an interior disposition, is the academic study of religion even possible?

If identifying the shortcomings of various approaches to definition prompts readers to throw their hands up in frustration – as if they were awaiting the correct definition to be revealed to them at the close of this book, much like learning that the butler did it by the end of a mystery novel – then they may have missed the point of the book's opening discussion of Douglas's work on classification. For instance, take the **cognitive scientist** and **linguist**, George Lakoff, who is well known for his work on classification. In a book entitled *Women, Fire, and Dangerous Things: What Categories Reveal about the Mind* (1987; the title of the book refers to the manner in which things are classified in traditional Dyirbal, one of the indigenous languages of Australia), Lakoff cites the philosopher of science, Stephen Jay Gould's discussion of the difficulties in settling on how best to classify that animal commonly known as a 'zebra'. Biologists, we learn, generally classify living things either *cladistically* (that is, cladists classify biological organisms in terms of their shared, and thus evolutionarily inherited, traits – somewhat like an essentilist) or *phenetically* (that is, pheneticists classify biological organisms in terms of their shared form, function and role – somewhat like a functionalist). Depending on which of these two different systems of classification the biologist uses, the three species of zebra end up being classified in rather different ways, with one of the three having more in common – as judged by the cladistic system – with horses than with the other two species of zebra.

The moral of this story of classification? If we presume, as Lakoff suggests, that there is one and only one proper way to classify the items of the world (a folk assumption that he thinks we commonly make in day-to-day life, but which might be rather mistaken), then clearly, when it comes to classifying those black and white striped creatures found in Africa, and in zoos throughout the world, we've got a big problem on our hands. For we now need to come up with a way to judge which classification system is the best to determine what zebras *really are*. But to do that, we need to employ yet another classification system, with internal criteria of its own, to judge 'better' from 'worse'. Moreover, if we understand language itself to be a classification system (i.e., What gets to count as a letter, a word, a sentence, a text? What gets to count as a correct meaning?), then even in speaking 'zebra' we find ourselves stuck in the midst of classifications systems not of our making, without which we might not be able to form a thought, much less get on with living our lives, get from point A to B, let alone be interested enough in striped animals so that we worry over how to classify them.

Perhaps, Lakoff says, the folk view that sees classification to be based in a singular **correspondence** between name and identity – an approach that assumes that zebras *really are* just one thing and thus that there is some definitive classification system capable of expressing it, if we could only find it – is itself the problem. What if, as Mary Douglas's work suggests, all we have are various groups of classifiers inventing and using a variety of systems that assist their members to achieve a variety of practical outcomes of importance to each group – much like coming up with a way to store books in a manner that makes them easily retrievable (should this be your interest) or grouping people in a manner that enables the distribution of goods (should this be your interest)? Instead of an essentialist view of classification, what if the systems we use to distinguish a this from a that are seen as practical human products, as inventions and tools that we make, use, fine tune, and, sometimes, discard (after all, few scholars today study 'the heathens', though scholars long ago took it for granted that there were such people and that they needed to be studied). If that was the case, then would the difficulties of each approach to defining religion prompt us to give up the search for a definition altogether? Or, instead, would these difficulties each be seen merely as the flaws that all human products inevitably have?

For those who think that religion names some essential thing existing out there in the world, or deep within the human heart – much like those who think that the things we call 'zebras' have some essential trait that *really* distinguishes them from the things we call 'horses' (which themselves have some trait that *really* distinguishes them from 'donkeys') – then the problem with the difficulties that we have identified with various attempts at definition is that none are perfect. But for those who see classification systems – such as **religion/culture** or **sacred/profane**, citizen/foreigner, familiar/strange, not to mention soil/dirt – as artifacts from human social worlds, artifacts that have a shelf-life and are bound to be obsolete some day, then these difficulties are not problems at all.

Due to the breadth of his own work and its international influence, **Jonathan Z. Smith** is, perhaps, the best known representative of a recent development among scholars of religion who take seriously that the category 'religion' – both the word and the various concepts that are attached to it – is *their* tool and that it does not necessarily identify a universal affectation lurking deep within **human nature**, and that there is therefore no one correct way to define it. Instead of siding with another famous Smith (**Wilfred Cantwell Smith**) and arguing that 'religion' is the sadly inadequate word outsiders use to name an inner **faith** that they can't really study in any sufficient manner (because we're studying it only by means of observing its various expressions, such as ritual, symbols, narratives, etc.), for J. Z. Smith 'religion' is understood as a linguistic, conceptual tool that some people happen to use in making sense of the worlds in which they find themselves; its definition is therefore intimately linked to their interests – whether their interests are intellectual and theoretical or whether they are practical and political. (Could one press this further and see all interests as practical? Marxist scholars think so, collapsing the common distinction between theory and practice, and calling it all praxis instead.)

In Smith's words, from the opening to his influential collection of essays, *Imagining Religion* (1982):

> While there is a staggering amount of data, phenomena, of human experiences and expressions that might be characterized in one culture or another, by one

criterion or another, as religious, there is no data for religion. Religion is solely the creation of the scholar's study. It is created for the scholar's analytic purposes by his imaginative acts of comparison and generalization. Religion has no existence apart from the academy. For this reason, the student of religion...must be relentlessly self-conscious. Indeed, this self-consciousness constitutes his primary expertise, his foremost object of study.

His point? As I read Smith he seems to be saying that, prior to exposure either to European missionaries, explorers or scholars, people the world over were not necessarily religious. That is, unlike the hypothetical newspaper editor, cited at the outset of this book, who had to decide if a story went into the religion or the local politics section of a newspaper, they did not use this particular term to group together and name specific items in their social worlds (certain beliefs, behaviors and institutions), in distinction from other elements of their social world. Moreover, historically speaking, because European Christians first came up with the term as a self-descriptor, they did not originally use it to talk about other people's beliefs, behaviors and institutions, for only they were religious and all others were heathens, sinners or superstitious. In its early uses, 'religion' was hardly a term that signified cross-cultural commonalities thought to be shared by all humans. To paraphrase Smith, 'religion' is a product of an interest in comparison and generalization brought about by means of an experience of how difference and similarity are interrelated.

In focusing attention on our category 'religion', we must therefore be careful not to presume that the word is merely the historical tip of a constant iceberg that defies history, for then we might find ourselves saying, as did earlier scholars (e.g., **Mircea Eliade**), that those untouched by European colonialism and its interest to distinguish the sacred from the profane, the religious from the political, lived in an idyllic world that was homogeneously sacred. Such a Romantic nostalgia strikes me as unproductive for it fails to examine the manner in which such peoples divided up and named their own worlds, so that they could come to know those worlds and act within them. Furthermore, this failure causes us to be uninterested in how other people establish their own relationships of similarity and difference akin to our soil/dirt or us/them. Sure, they may have assumed that non-human **agents** could affect their worlds, they may have performed behaviors of relevance to these beliefs, and yes, they likely had views on how the universe came about and where it was going – but they did not call these things religion and, most importantly, they did not necessarily link these beliefs and behaviors together, seeing them all as somehow exhibiting an identity that made them separable from a host of other daily, mundane actions necessary for life to go on. So, calling those things 'religious', 'holy' or 'sacred', in distinction from

other sorts of things, may well be *our* act of classification, how we make
the world knowable to us – an act performed by curious outsiders who,
by means of sets of tools and concepts they inherit from their teachers,
set about to satisfy curiosities about the relationships (or lack of) between
themselves and others. Like the magnifying glass on the cover of this
book, the classification 'religion' – both the word and the many different
things it can mean – may be one of the lenses we commonly use to know
our generic surroundings as a world comprised of significant relationships,
identities and values. In both the case of 'religion' and magnifying glasses,
such devices can be very useful, all depending on our goals, but we'd be
terribly mistaken to think that the world as we come to know it through
each is the world as it really is.

Therefore, contrary to **Max Weber** (1864–1920), who famously opened
his now classic book, *The Sociology of Religion* (1922), by stating that
exhaustive description must come before any attempt to define religion –
'Thus the final and definitive concept cannot stand at the beginning of
the investigation, but must come at the end' – some scholars no longer
see classification to be concerned with linking an **historical** and therefore
material word to an ahistorical and purely ideal quality identified only
after all empirical cases have been considered. Instead, classification –
like all human activities – is now understood by many as a tactical, provisional activity, directed not by **inductive** observation followed by generalization but, instead, by **deductive** scholarly theories and prior social
interests that are in need of constant disclosure and close examination. In
this way, classification ensures that some generic thing (such as what
Mary Douglas termed 'matter' [a word that seems to presuppose an atomic
view of reality], what we might more informally call 'stuff', or what I just
referred to as our generic surroundings) stands out as an object worthy of
our attention; come to think of it, without a prior definition of religion
Weber would not have had anything to describe, for how would he have
known what to include in his investigation? To paraphrase Jonathan Z.
Smith, writing in the Afterword, classification therefore provides scholars
with some elbow room to get on with the work of disciplined inquiry
which is itself prompted by their curiosity concerning how the world does
and does not conform to their expectations.

It is therefore fitting to end this brief introduction to some of the tools
used when studying religion with the words of the scholar of **Hinduism**,
Brian K. Smith, who, in his book *Classifying the Universe: The Ancient
Indian Varna System and the Origins of Caste* (1994), offers a rather different view of definition from that of Weber – one that is nicely in line with
that of J. Z. Smith: 'To define', he writes,

is not to finish, but to start. To define is not to confine but to create something and...eventually redefine. To define, finally, is not to destroy but to construct for the purpose of useful reflection... In fact, we have definitions, hazy and inarticulate as they might be, for every object about which we know something... Let us, then, define our concept of definition as a tentative classification of a phenomenon which allows us to begin an analysis of the phenomenon so defined.

Classification systems, then, do not have to be seen as illuminating some deep, essential or necessary trait – whether we are defining mountains, soil, zebras or religions. Instead, they are our own tactical and always provisional tools that provide us with a little wiggle room so that we can get on with the production of knowledge and action in the world. After all, as pointed out time and again by J. Z. Smith, because all knowledge is based upon classification systems, we ought to be interested in where our ordinary classifications come from and how they work; for, as he states in the conclusion to his 2000 essay entitled 'Classification' (published in the *Guide to the Study of Religion* [2000]), 'the rejection of classificatory interest is...a rejection of thought'.

So, to answer the question posed at the outset of this book – 'What is the study of religion?' – we can now say that, at least for some scholars, it is the disciplined inquiry of but one aspect of human cultural practices – an aspect identified, for the purposes of our study, by the definition we as scholars choose to use, a definition that suits our purposes and our curiosities. What unites us into this collective group – signified by the possessive pronoun 'our', as in 'our purposes' – is not only our shared curiosities, common tools, and agreed-upon standards of argumentation, but also the common institutional setting that draws us together, and to which our labors contribute. This setting is, for many of us, the public research university, an institution that has profound bearing on what ends up counting as the academic studying of religion.

Perhaps, then, we should conclude by revising our original question, for 'What is the study of religion?' might best be answered by first asking, 'Where is the study of religion being practiced, by whom, and for what purposes?' For, depending on its context and the interests that drive it, the study of religion can be very different things – something like the generic stuff of the world just as easily being understood as either soil or dirt.

Afterword

If concepts, categories, and systems of classification are tools that we devise and use for purposes, then what about such things as a college course's syllabus or a book such as this? Would they not also be historical artifacts rather than things that spring from the ground overnight, like a mushroom? Taking the study of human behavior seriously means understanding scholarship itself to be but one more human practice, yet another way of trying to make the world knowable.

Afterword: The Necessary Lie: Duplicity in the Disciplines

Jonathan Z. Smith[*]

George Bernard Shaw once made a wisecrack that I think defines the academic disciplines as social entities: 'I may be doing it wrong but I'm doing it in the proper and customary manner'. This raises at least two questions that I would like to examine. First is the white lie, which comes up when we are self-conscious about speaking in a nondisciplinary fashion about our subject. Second is disciplinary lying, which is part of the process of initiating somebody into a discipline. Indeed, disciplinary lying may be the marker of what it is to belong to a discipline.

The White Lie

We lie, it seems to me, in a number of ways. We sometimes cheerfully call the lie words such as 'generalization' or 'simplification', but that's not really what we're doing. We're really lying, and lying in a relatively deep fashion, when we consistently disguise, in our introductory courses, what is problematic about our work. For example, we traditionally screen from our students the hard work that results in the production of exemplary texts, which we treat as found objects. We hide consistently the immense editorial efforts that have conjecturally established so many of the texts we routinely present to our students as classics, not to speak of the labors of translation that enable many of them to read these texts. Then we read them with our students as if each word were directly revelatory, regardless of the fact that the majority of the words are not in the

* This brief address was originally presented at the University of Chicago's Center for Learning and Teaching and was subsequently included in the pamphlet, *Teaching at Chicago: A Collection of Readings and Practical Advice for Beginning Teachers*, Diane M. Enerson (ed.). Chicago Teaching Program, University of Chicago. It is reproduced here with the permission of the author.

language in which the text was written. In fact, we have a curious strategy of when and how we decide to display some of this hard work. For example, Chinese or Japanese texts in translation read like Yiddish – every third word is followed by some indecipherable foreign word in parentheses as if this would in some way enhance understanding. We are really reminding our students that this is foreign and hard to understand. In Shakespeare, we display an enormous glossary material, implying that this, too, is a foreign language that, nevertheless, can be mastered with effort. Yet the King James Bible, another Elizabethan text, is characteristically taught in God knows how many humanities courses across the country with never a single footnote indicating that the language, while simpler than the language of Shakespeare, is just as foreign and just as difficult. One would like them to note, for example, that the word 'let' often means to stop somebody from doing something, and the word 'prevent' at times means to let them go ahead and do it. One gets odd moral conclusions by reading the King James Bible without such footnotes, and yet our mutual lie is that it is infinitely accessible while Shakespeare is accessible with difficulty; foreign texts remain inaccessible.

Moreover, we conceal from our students the fields-specific, time-bound judgments that make objects exemplary. We display them as if they are self-evidently significant and allow the students to feel guilty when they do not feel this self-evidence. We rarely do what some German critics have called a reception history of the object in front of us, examining why or how the object became in some way exemplary of humankind in a particular discipline. Thus, when we deal with a figure like Plato, we rarely reflect on the fact that, after all, the dialogue that was Plato for the Western world for most of its history (i.e., *The Timaeus*) is no longer read. Jefferson and other wise people despised *The Republic* thoroughly, finding it an absolutely impenetrable document. They thought Cicero – today all but dropped from the canon – was the place one went in order to think about democratic institutions. That is, we don't introduce our students to the fact that the artifacts that we examine are scarcely blooming with self-evidence. We conceal the revisionary histories of the objects we examine. If they're written works, we conceal their drafting and their changes. If they're scientific objects, we conceal the history of failed experiments and the history of sheer serendipity. That is to say, we convey to our students a specious perfection of the object and a specious necessity to the history of that object.

When we conceal from our students our hard work, that which is actually the way we earn our bread and butter, we produce a number of consequences. I remember testifying once before the California state legislature and facing a legislator who wanted to know why professors should be paid to read novels, when the legislator himself read novels on the train every day. Well, that was the price of our disguising the work that goes into things. There are, I think, more serious educational consequences. If we present the work as perfect or as work without a revisionary history, then we present a work that no student could hope to emulate. Indeed it serves, if it serves at all, as a standard for how far below that standard the student falls. If we present the material without displaying the effort that goes with it, students tend to conclude that things are true or false, or alternatively, that it's entirely a matter of their opinion whether the object is exemplary. In that case, what we have is a contrast between his or her feelings and my feelings. Thus, in the name of simplification, what we really end up doing is mystifying the objects we teach at the introductory level.

Similarly, still in the name of simplification, we treat theory as if it were fact. We treat difficult, complex, controversial, theoretical entities as if they were self-evident parts of the universe that we inhabit. Students coming out of introductory courses in the humanities know that there is such a thing as an author's intention, and they regularly and effortlessly recover it from the text they are looking at. Students in introductory social sciences know that there is such a thing as a society that functions, and they effortlessly observe it doing so. Students in introductory sciences are wedded without their knowing it to a tradition of induction from naked facts, in what Nietzsche called 'the myth of the immaculate perception'. Indeed, I've often argued when teaching in the social science Core that, if I could only have the first week of Chemistry 101, my job would be infinitely easier because at least we would have raised the possibility that one wears eyeglasses when one gazes at these naked facts.

Despite the proud claim that we make over and over again that we teach the how rather than the what of the disciplines, we, in fact, do not; it is the theoretical conclusion that our students underline in their books. I spend a half hour with each of my students looking at what they've underlined, and they've always underlined the punchline and never anything that might be called the process that led up to it. That is to say, theoretical entities have been reduced to naked facts. The process of discussion often becomes one of show and tell for these unproblematic, now self-evident conclusions. In other words, we have skillfully concealed from our students the power of the remark once made by a mathematician, 'I have my results, but I do not know yet how I am to arrive at them'. Even a false generosity with respect to method conceals the process when we present this method one week, that method another week, allowing none of them to have the kind of monomaniacal power or imperialism that a good method has when we're honest about it. Without the experience of riding hell bent for leather on one's presuppositions, one is allowed to feel that methods have really no consequences and no entailments. Since none of them is ever allowed to have any power, none of them is ever subjected to any interesting cost accounting.

Another way we end up reducing our students to the notion of a subject being all opinion (and we're very angry when they assert that to us) is the way that introductory courses, whether seminar or lecture, whether of a large field of study or a small field of study, are never introductions. They are always surveys. They may be shorter surveys or longer surveys, quicker surveys or slower surveys, but nothing is allowed to be truly troublesome. It suggests that one might think that a freshman seminar devoted to a single work is probably a far better introduction than our vaunted Core. That is to say, one really ought to be able to work on a limited number of exemplary objects and to answer all the various sorts of questions that one might come up with. Though I don't like a lot of the framework, Jeff Robinson has a book, *Radical Literary Education*, about a classroom experiment in which he takes the introductory English class through a reading of a Wordsworth ode for an entire semester at Colorado State. They're into a complex unpacking and unfolding of the enterprise. I'm not terribly thrilled with the message he'd like you to get from this; nonetheless, the strategy, it seems to me, is one worth looking at.

Disciplinary Lying

The self-justified white lie is done in the name of our students, in the name of simplifying, of generalizing, of speaking to a wide and a diverse audience. However, one also has to look at the place in which lying becomes built into the structure of things, in which it becomes that which constitutes a discipline as a discipline over and against other disciplines. Here, at least in principle, we lose the excuses that go with the introductory course. One would presume a student who had been through a program of rigorous disciplinary lying would emerge at the conclusion of his or her baccalaureate experience with some measure of sophistication. Yet, when I used to do something called the dean's seminar in which we talked about the disciplines as seniors graduated, I was struck by their lack of the sense of the conventionality that governs what we do. These seniors still sought the cost-less method, the cost-less theory, even at the end of two to three years of allegedly depth study in a field.

Fields are taken not only as self-evident but as singular, without real understanding that what's a style for one is not a style for another. Take a simple example in my own field. If I want to publish an article in one of two general journals in the field of religion – *History of Religion* and *The Journal of the American Academy of Religion* – I have to at least redo the notes. *History of Religion* does the so-called humanities-style notes and *The Journal of the American Academy of Religion* does the so-called social science-style notes. It's not just that it's inconvenient; what I am doing is fundamentally altered by which of those two styles I accept. In the humanities, the footnote is exegetical, and you will accept what I say on the basis of my exegesis of that particular passage. On the other hand, when I read something that says, 'Levi-Strauss 1970: 83', I'm supposed to find the one sentence in a four-volume work that justifies the paragraph I have just read. That's a very different understanding of how you justify your work. That really is an authority model, which has very little to do with any claim to exegesis. Yet, one never talks about such differences with students.

I discovered a stunning example of disciplinary lying in a book by the now late Nobel Prize winner, Richard Feynman, *Surely You Are Joking, Mr. Feynman*, written for no other purpose that I can determine but to make money. He writes, rather cockily, that he finds world travel a rather dull way of spending a vacation, so instead he travels to another discipline. He spent one summer working in the biology laboratories at Cal Tech, and, according to his report, his results were significant enough to interest James Watson and have him invited to give a set of seminars to biologists at Harvard. Yet when he wrote up his results and sent them to a friend in biology, his friend laughed at Feynman. As he recalls, 'It wasn't in the standard form that biologists use, first procedures and so forth. I spent a lot of time explaining things that all the biologists knew. Edgar made a shortened version, but now I couldn't understand it. I don't think they ever published it. I learned a lot of things in biology. I got better at pronouncing the words, knowing what not to include in a paper or seminar and detecting weak technique in an experiment'.

Now, that's really, when you stop to think about it, a rather remarkable paragraph. Consider how much Feynman is signaling when he uses the phrase, 'It wasn't in the standard form that biologists use'. Feynman tells us that he did get some sense of the language domain of the field – how to pronounce the words; he did learn something of the tacit conventions – what not to say, what was not needed to say; he learned something about what counted as appropriate according to the conventions of the fields. What he could not recognize was the fictive modes of accepted disciplinary discourse. As a result, we have a Nobel Prize-winning physicist who, when he writes up an experiment, is laughed at by his biological colleagues; when they write it up 'properly', he is incapable of understanding his own work.

This is what lying in the disciplines is all about. It is constructed very much as an initiatory process. As some of you may know, among the southwestern Amerindians, as well as among a number of other people, initiation consists of an act of unmasking. Certain figures wear masks and are called gods. When you reach the age of maturity, the elders take you to the other side, the figures take off their masks and show you, 'hah, hah, hah, it's just good old Uncle Joe', as if you hadn't recognized that earlier. At least the convention is 'now we unmask'. A great deal of what a discipline does is initiating its neophytes, pulling rugs out from under things you thought you knew and unmasking things you thought were clear. The initiated use another kind of language, forming a set of those who are in on the joke.

When we talk about disciplinary instruction, we're talking about creating a corporate entity arrived at through an initiation that proceeds through a rigorous sequence. Within some of the sciences, in theory at least, that sequence is carefully arranged. It's carefully structured from elementary school to postdoctoral work as one endless and lengthy series of unmasking what you thought you knew. The ideal, often quoted in books on science and education, is the breathless individual who, when Oppenheimer was at the Institute for Advanced Research at Princeton, was asked, 'What is it like to study with Oppenheimer?' and who responded, 'It's wonderful. Everything we knew about physics last week isn't true'. Well, this is what it means to be an initiated member of a discipline. The science you learned in elementary school is no good when you get to high school, which is no good when you get to the first year in college, which is no good by the second year of college, and so forth.

What, however, happens to the person who doesn't stay the course? This notion of the delayed payoff is problematic. My son came home very depressed from high school chemistry because he said he 'got an experiment wrong'. I told him that you can't get an experiment wrong. An experiment is trying to find out something. You put your two things together, and you found out something. He said, 'No, no, no, it wasn't the way it was supposed to come out'. Well, then it wasn't an experiment. If he performed the same experiment in college, they could show him twenty-eight more variables that went into the results, and he would have understood that he didn't get it wrong. If that is his only experience with science, he'll never have that particular idea unmasked.

In most of the fields that we teach, there is no such even rudimentary recognition of sequence or corporate responsibility. Too often the sequence listed in the course catalogue is only political, requiring one course with each professor in a department. The majority of concentration programs, or for that matter graduate programs, don't acknowledge the underlying initiatory sense that what we knew for sure yesterday we now know as somewhat problematic.

Though I think there is something to disciplinary lying, I think there is very little to justify introductory lying. In the case of the introductory courses, we produce incredibly mysterious objects because the students have not seen the legerdemain by which the object has appeared. The students sense that they are not in on the joke, that there is something that they don't get, so they reduce the experience to 'Well, it's his or her opinion'. On the other hand, disciplinary lying – the conventions within a discipline – enables me to get moving. You have to allow me some measure of monomania if I am to get anywhere. I can't do my work when I have to stop and entertain every other opinion under the sun. This is why such work must always be done in a corporate setting, so that the monomania's mutually abrade against, so that they relativize each other; so that the students, the initiates, are let in on the joke. I had an old teacher who, when you said something you thought was very smart, would say, 'That's an exaggeration in the direction of truth'. I have always thought that was the best definition I have ever heard of the academic enterprise.

Glossary

Each time a technical term is first used in any of the preceding chapters it is printed in bolded text, signaling that it is defined and discussed in greater detail in this section of the book. In addition, within each of the following definitions other technical terms are printed in bolded text, since they too are defined in this alphabetically arranged section. Finally, the names of relevant scholars whose works are discussed within this definition section are also bolded; discussions of their works are found in the next section of the book.

Agent – term commonly used to refer to a being assumed to be **intentional**, that is, a being who acts, has motivations that inspires such actions, and can therefore be held accountable for these motives and actions. Human beings are therefore thought to possess the quality known as agency. The term is also sometimes used to describe non-intentional things, such as a 'chemical agent', which nonetheless are thought to be able to cause certain outcomes. A traditional way of defining religion is that it is a system in which agency is attributed to super-human powers (e.g., gods, ancestors, etc.).

Agnosticism – term coined in the nineteenth century by combining the Greek *gnosis* (meaning esoteric or secret forms of knowledge) with the prefix a- which often denotes the negative form of a word; a philosophical position that admits to having no privileged knowledge concerning whether God or the gods exist; a position of theological neutrality to be distinguished from **atheism** and **theism**. Methodological agnosticism is the name given to the neutral position some scholars of religion argue one ought to take. This implies that, regardless of one's personal viewpoint, as a scholar one employs tools (i.e., methods) that avoid asking questions of truth.

Animism – [Latin *anima*, meaning life, soul] a term popularized by the late nineteenth-century anthropologist **Edward Burnett Tylor** to name the belief he thought to be held by **evolutionarily** early people (what Tylor would have named as 'primitive' or 'tribes very low in the scale of humanity') concerning natural phenomena (e.g., trees, the ocean, people, etc.) possessing spirits or souls. This term, and his **theory** of animism, was developed to help answer the question: 'What is the origin of religion?' making Tylor an early example of a scholar developing a **naturalistic** theory of religion.

Anthropology – [Greek *anthropos*, meaning human being + Greek *logos*, meaning the systematic study of] the modern, comparative and cross-cultural science that deals with the origins, physical and cultural development, biological characteristics, and social customs and beliefs of humankind. Practiced as a component of the **human sciences**, the academic study of religion is considered distinct from the discipline known as Anthropology though Religious Studies (as it is known in North America) could be said to be anthropological in its outlook (or what is sometimes termed 'anthropocentric': centered on the study of human behavior); that is, when practiced as something other than **theology**, the study of religion is focused on human beings and their practices and does not study the gods and their will; see **Human Sciences**.

Anthropomorphism – [Greek *anthropos*, meaning human being + Greek *morphé*, meaning shape or form] as in personification, to ascribe a human form or human qualities and traits to non-human things; prosopopoeia [from the Greek, *prosopopoiia*, to make a mask or face] is a related term, naming the poetic technique of having a dead or imaginary person speak, as well as the technique of giving human qualities to inanimate objects such as mountains or the sea. 'The sea was angry' could be considered an anthropomorphic claim; seeing faces in the moon, or faces in the patterns found in wood grains, could also be considered evidence of anthropomophism. Central to **David Hume**'s early **theory** of religion, a **modern** theory of anthropomorphism is that of the **anthropologist** and **cognitivist**, Stewart Guthrie, who argues in his book, *Faces in the Clouds: A New Theory of Religion* (1993), that humans – among many other species – possess brains that are 'hard wired' to project onto the world the traits that they perceive themselves to possess, all in an effort to make sense of, and thereby navigate, an otherwise un-known environment. For Guthrie, much as with Hume, religion (the widespread belief that the universe is a living **agent** that cares for human beings) is but one instance of this anthropomorphic strategy.

Atheism – a term that combines the Greek *theos*, meaning god + the negative prefix a- which often denotes the negative form of a word; the philosophical position that denies the existence of God or the gods; to be distinguished from **theism** and **agnosticism**.

BCE/CE – Unlike the explicitly **Christian** calendrical system known to most people in North America and Europe and which is based on the Gregorian calendar – with BC standing for 'Before Christ' (in the English version, based on the older ACN, which is Latin for *Anti Christi Natus*, meaning 'before the birth of Christ') and AD (stand-ing for the Latin phrase *Anno Domini*, 'year of Lord', short for *Anni Domini Nostri Jesu Christi*, meaning 'In the year of our Lord Jesus Christ') – BCE and CE are favored by some scholars, especially those who study other cultures which have their own entirely dif-ferent calendrical/dating systems (e.g., the **Islamic** calendar begins in what readers of this book might call the year 622, when the first fully integrated Muslim community was founded in the Arabian city of Medina). Although this alternate dating system uses precisely the same numbers, their initials stand for 'Before the Common Era' and 'Common Era', referring to the adoption of a common calendar in the ancient Greco-Roman world. Scholars often adopt this alternative notation to avoid the explicitly **theological** assumptions of the so-called Western dating system, which roughly revolves around the time when Christians traditionally believe Jesus to have been born.

Buddhism – the name given to a collection of beliefs, practices and institutions that developed from (sometimes said to be in reaction to) **Hindu**/Indian institutions and that revolve around the importance placed upon the teachings attributed to Siddhartha Gautama, thought to have lived and taught in northwestern India between the sixth and fifth centuries **BCE**. Gautama is known by the honorary title of 'the Buddha' (which, in the language of Pali, means 'awakened one'). The Buddha is said to have awoken to the true nature of reality, thereby experiencing *nirvana* (to extinguish one's presumption of having a distinct, enduring self). His teachings involve understanding that all appearances are misleading and that impermanence, or change, is the basis of all reality. Several dominant branches of Buddhism exist today and it has distinctive shape in different geographic locations (such as in southeast Asia as opposed to Tibet, China, Japan, Europe and North America). Studies of Buddhism will often begin by narrating the life of Gautama (given that it illustrates certain key ideas that come to symbolize basic Buddhist doctrines), and then focus on its critique of Hinduism's *caste* system as well as the doctrines known as the Four Noble Truths (credited to Gautama's first teaching after attaining enlightenment) and the Noble Eightfold Path (entailing a systematic behavioral system of detachment or mindfulness). Although 'Buddhism' is an outsider's term (coined under the earlier European pre-sumption that this Asian mass movement is centered on the worship of the Buddha, just as **Christianity** was understood by them to be centered on the worship of the Christ), a more apt term for this tradition may be 'the Middle Path' (between the two extremes of craving and complete renunciation).

CE – see **BCE/CE**.

Christianity – the name given to a collection of beliefs, practices and institutions that developed from out of the ancient **Jewish**, as well as the Greco-Roman, world of antiquity. Focused on the life and teachings of a turn-of-the-era Jew named Jesus of Nazareth, it began as an oppositional movement that was persecuted and, by the early fourth century **CE**, it had become tolerated throughout the Roman empire. Its teachings, found in its scripture called the Bible (from the Greek for paper, scroll or book), include much of the previously existing Jewish scripture, including the Torah, along with the New Testament comprising the Gospels (from the Greek for 'good news'), which present various narrations of the life and significance of Jesus (including his resurrection from the dead after being executed by the Roman authorities), along with the Epistles (Latin *epistola*, meaning letter), comprising communications between early Christian leaders (such as the influential early converts from Judaism to Christianity and its most important early missionary, Paul [or Saul] of Tarsus) and various isolated early Christian communities or house churches. Jesus, considered early on to be the *messiah* ('annointed one of the Lord', a Hebrew designation originally of relevance to Jewish tradition) was soon understood by his followers to have been 'the son of God', and later in Christian doctrine is understood to have been one of three aspects of God (the trinity, also including God the Father and the Holy Spirit). The honorary title of 'Christ' (from *khristos*) derives from the **Greek** translation of the Hebrew *mashiah*; Christians are therefore followers of the one believed to be the Messiah. Currently, Christianity involves three major sub-types, some of which differ significantly from the others on issues of doctrine and ritual: Roman Catholicism, Protestantism (which contains a large number of sub-types), and Greek Orthodoxy.

Church/State – a dichotomy whose origins date to about seventeenth-century Europe, commonly used today in the United States to stand for the legally mandated separation between the workings of any Church or religious group and the State; this notion of separation is traced to the **First Amendment** of the US Constitution. Commonly, US political theorists and legal scholars refer to the 'wall of separation' between Church and State, although this widely used phrasing is not in the Constitution. Instead, it derives from phrasing found in a letter written by Thomas Jefferson (1743–1826) while he served as the third President of the United States (1801–1809).

Cognitive Science – the systematic study of the precise nature of different mental tasks and forms of cognition, and the operations of the mind/brain; this study uses elements of **Psychology**, Computer Science, Philosophy, and **Linguistics**. In recent years, it has proved one of the more active and organized sub-specialties in the academic study of religion, focusing specifically on the study of **ritual**.

Unlike some popular forays into the interface between religion and cognitive studies, such scholarly work seeks not to isolate the part of the brain that experiences God or the **sacred** (such as the so-called 'god gene'); instead, as in the work of **Pascal Boyer**, they apply findings from Cognitive Psychology to develop a **naturalistic theory** of religious beliefs and behaviors.

Colonialism – the economic or political control or governing influence of one **nation-state** over another (a dependent country, territory, or people); also, the extension of a nation's sovereignty over another outside of its boundaries to facilitate economic domination over the latter's resources and labor usually to the benefit of the controlling country. Although not limited to European nations, the rapid expansion of their influence all across the globe during the eighteenth and nineteenth centuries today attracts a great deal of attention among scholars and has led to the development of a new field known as Post-colonial Studies, which focuses on the implications of, and local reactions to, the colonial era.

Comparative Religion – a systematic study of the commonalties and differences among the religions of the world; this study seeks to establish a set of principles and categories that can be used systematically to understand the universal and particular features of religions (in the plural) and to determine whether they are sub-types of **religion** (in the singular). Although today the notion of comparative religion is sometimes limited to the work carried out in a **world religions** course and textbooks, during the field's formative period in the late nineteenth century (especially in Britain), Comparative Religion was the name often given to the entire field. See **History of Religions**.

Confucianism – name given by European scholars to a group of Chinese schools of thought associated with the teachings of such writers as Confucius (551–479 **BCE**), Mencius (372–289 BCE), and Hsun-tzu (298–238 BCE). These traditions focus upon developing proper forms of social and political behavior. During the Chinese Han dynasty (206–220 CE), these schools became official sate orthodoxy, and a authoritative collection of texts and temples were established; see *Li*.

Contingent – see **Necessary**.

Correspondence Theory – a common approach to understanding how truth and meaning-making works, and thus how definition works; the truth of some claim (or, what we might better call a proposition, such as 'The sky is blue') is thought to be determined by whether or not the claim fits, or corresponds, to some observable set of facts. The truth of language (for example, the words strung together in a sentence) is therefore thought to have a direct relationship with an observable, stable reality and the judgment 'true' is therefore a confirmation of this relationship. This correspondence theory (also called a referential theory insomuch as words are thought to gain their meaning insomuch as they refer to real things in the world) applies equally well to the production of meaning, since it is commonly thought that a word – say, the word 'blue', as in the proposition 'The sky is blue' – refers or corresponds to some quality a thing possesses – in this case, the quality of blueness possessed by the sky. Or, to phrase it another way, among the sky's many observable characteristics is one that is of particular interest to us right now, its color. I look up, confirm it is blue and thus judge the proposition 'The sky is blue' to be true and the word 'blue' to be meaningful; see **Positivism**.

Cult – [Latin *cultus*, meaning care, cultivation, and by extension, a system of **ritual**] originally a merely descriptive term for the ritual component attached to any social group, as in the phrase 'the cult of the saints' (implying routines of Roman Catholic devotion focused on **Christian** saints), it is today a term most often used today in popular **culture** to name marginal groups considered by members of dominant groups to be deviant and thus dangerous (somewhat akin to the pejorative term 'fanatic' and 'fanaticism'). In the **sociology** of religion, 'cult' is classically used as a technical term, in distinction from both 'church' (or 'denomination') and 'sect', to signify differing groups' varying degrees of social integration. Traced to the work of the German sociologist, **Max Weber**, 'church' and 'sect' were technical terms he used to identify what he took to be significant differences among religions, the former meaning a religion into which one was born whereas the latter named one in which membership was the result of a conscious decision. This pair of terms was then reformulated by the German theologian, Ernst Troeltsch (1865–1923) – such as his book, *The Social Teaching of the Christian Churches* – 'church' was distinguished from 'sect' in terms of the latter being a group in greater tension to the dominant social world whereas the former being a group that more easily accommodates itself and, thereby, lives in greater harmony with the wider social world. For Troeltsch, 'mysticism' was the term he used for a third, far more private and individualized variation that likely did not lead to any form of social organization. In the early 1930s, the sociologist Howard Becker termed this latter group 'cult'. The **modern**, popular use of the term to name groups that deviate too far from accepted conventions can be understood to develop from these uses.

Culture – [adaptation of Latin *cultura*, meaning cultivation, to tend, hence involving the notion of domestication] that portion of thought and behavior used by social groups that is learned and capable of being taught to others; culture can include: language, customs, **worldviews**, moral/ethical values, and religions. For those who believe that religion (or at least some elements of it, such as so-called religious **experience**) is somehow set apart from all other aspects of the historical world (making religion *sui generis*), the concept of culture is sometimes set apart from that of religion and they are thought to interact in specific ways – hence the 'and' that joins the popular designation 'culture and religion', a phrase found in the work of many scholars of religion. For yet others who are more **anthropologically** inclined, and who therefore see religious practices as but a sub-set of cultural behaviors, that can be explained in precisely the same manner as all other cultural attributes, it would be more appropriate to talk about 'religion in culture'.

Deduction – a process of reasoning in which the conclusion follows necessarily from the premises presented so that the conclusion cannot be false if the premises are true. Deductive logic is a form of argumentation in which one begins with acknowledged general premises and then reasons from these to specific cases, such as the three-part form of reasoning known as the syllogism in which the major premise (All people are mortal) is followed by a minor premise (Judy is a person), which logically leads to a specific conclusion (Judy is mortal). The role of inductive reasoning to establish the major premise of a syllogism ensures that deduction and **induction** are intimately connected forms of reasoning.

Dialectic – traditionally understood as the question/answer teaching style used by Plato's character Socrates; in later European philosophy it stands for a progressive series in which the opposition between a thesis statement ('The sky is blue') and its opposite, the antithesis ('The sky is not blue'), is resolved by means of a synthesis ('The sky is partially blue') which itself comes to be understood as but a new proposition (i.e., thesis) which has its opposite that can again be resolved by means of a synthesis.

Din – Arabic term (pronounced 'deen') found in **Islam** that is often translated into English as 'religion'. It is thought that the term dates to a
much earlier idea of an actual debt that must be settled on a specific date, which in turn led to such other meanings and usages as:
the idea of following an established series of customs for settling
debts; the act of guiding someone in a prescribed direction to carry
out required action; the act of judging whether such a prescription
has been followed properly; and, finally, visiting retribution upon
one who has failed to follow the required prescriptions. If this etymology is persuasive, then the link from the earlier notion of an
actual debt to the later notion of the manner in which Allah judges
human beings can be understood as a rather sensible development
of the concept. See also *Eusebia*, *Li* and *Pietas*.

Discourse – most simply, the communication of thought by words/con-
versation; a discourse could therefore be likened to a conversation
or, more technically, to a teaching or a systematic exploration of a
topic; many scholars now use the term to refer to any number of
fields or disciplines, the formal discussion of a subject in speech or
writing, or, following the French **postmodern** scholar, Michel
Foucault (1926–1984), even the series of **material** as well as intellectual conditions, practices, institutions, architecture and conventions that make specific types of thought and action possible (such
as the discourse of the academy or the discourse of medicine).

Emic and Etic – terms derived from the suffixes of the words 'phonemic'
and 'phonetic'; the former refers to any unit of sound significant to
the users of a particular language (each such unit of sound is known
by scholar of linguistics as a phoneme) and the latter refers to the
system of cross-culturally useful notations that represent each of
these vocal sounds (as in the phonetic alphabet found in the front
of most dictionaries and used as a pronunciation guide); derived
from the same **Greek** root, 'phonemic' designates the complex
sounds themselves whereas 'phonetic' specifies the signs and systems scholars devise to represent and then compare the manner in
which the basic phonemic units of a language are produced and
pronounced. Adopted by **anthropologists**, and later by scholars of
religion, the terms emic and etic come to stand for the participant's
(emic) and the non-participant's (etic) viewpoints.

Empirical – term used to name something that can be observed or perceived with one of the five senses: taste, touch, smell, sight and
hearing.

Essentialism – an approach to definition that maintains that membership within a class or group is based on possessing a finite list of characteristics or traits, all of which an entity must **necessarily** possess to be considered a member of the group, as opposed to the merely accidental or **contingent** characteristics a thing might or might not possess (sometimes also known as the substantive approach). An essentialist view of religion asserts that there are many different characteristics to be found among religions, but argues that these characteristics are merely secondary and superficial; instead, there are a small number of primary characteristics, possibly only one (its so-called essence or substance), that encompass all the religions of the world within one category; see **Existentialism**, **Family Resemblance** and **Functionalism**.

Establishment Clause – a clause contained in the **First Amendment** of the US Constitution that prohibits Congress from 'respecting an establishment of religion'. Many read this clause as meaning that Congress is not allowed to create a national religion, give preference to one religion over another, or prefer a religious over a secular outlook, but others argue that there is ambiguity in the clause itself concerning its use and implementation.

Eusebia – ancient Greek term for the quality one was thought to possess if one properly negotiated the various social expectations and duties required based upon such factors as one's social rank, gender, birth order, generation, occupation, etc. Often translated as 'piety' (from the Latin *pietas*), it is not to be confused with the **modern** sense of '**religion**', insomuch as the quality of *eusebia* resulted from one's proper behaviors toward the gods (such as performing a **ritual** in the prescribed manner at the appropriate time and place) but also from those behaviors involving one's social superiors, equals and inferiors. Therefore, piety in the Greco-Roman world was a fundamentally social, and not a **faith**, designation. See *Din* and *Pietas*.

Evolution – **theory** developed in the nineteenth century by such scholars as **Herbert Spencer** and Charles Darwin (1809–1882) to explain biological change in a population from generation to generation by such processes as random mutation, natural selection and genetic drift. The much criticized theory known in the nineteenth century as Social Darwinism names a school of thought that applied this biological theory to account for cultural and racial changes over time and place (assuming a uniform, linear development from so-called lower or primitive cultures to so-called higher or civilized cultures). Today, teaching evolutionary biological theory in public schools is controversial in some areas of the US due to the manner in which it is understood by some **Christians** to contradict a literal reading of the creation of the world as found in the Bible's book of Genesis. Although so-called Creationism, Creation Science, and what is now known as Intelligent Design, have all been proposed as an alternative to evolutionary theory, and in some cases are taught alongside it in public schools, so far no non-Christian views on the creation of the universe (such as the **Hindu** view whereby the god Brahma periodically creates a new universe from raw material that remains after the god Shiva destroys the previous one, which had decayed to the point of utter corruption) have gained sufficient support in the US to prompt them also to be taught in the public school system as competitors to evolutionary theory.

Existentialism – although it can be traced to earlier influences, it is primarily understood today as a mid-twentieth- century European philosophical movement, much associated with post-World War II French intellectuals (philosophers, literary critics, authors, playwrights, etc.), that takes as its starting point the priority of the individual along with the assumption that, in the words of one of the best known representatives of the movement, Jean Paul Sartre (1905–1980), 'existence precedes essence' – that is, **historical** human beings come before, and are thus the makers of, qualities and values. As Sartre also observes, human beings are therefore 'condemned to be free' – that is, as **agents** they have no choice but to be accountable for their own actions, desires, and the values they produce. Existentialism, then, can be understood to be in opposition to **essentialist** approaches to the study of **culture** and meaning, though there have been **theological** existentialists.

Experience – many **humanistic** scholars of religion argue that religion is grounded in a unique type of experience, conceived as an inner, personal sentiment that can only be expressed publicly by means of symbolic actions (e.g., language, **ritual**, etc.) that are themselves derivative and thus flawed copies of the original (a position represented by the work of **William James**). As made evident by the British literary critic, Raymond Williams, in his book *Keywords: A Vocabulary of Culture and Society* (1976), there are two senses of the term experience, distinguishable in English literature from around the late eighteenth century: historically related to the word 'experiment', its first sense can denote the accumulation of empirical facts and the results of such an accumulation, such as one having 'work experience' (Williams terms this sense 'experience past'); the other, which he describes as 'experience present', denotes a form of ever present consciousness that resides within the individual and to which one appeals when making judgements concerning the authenticity of a person. It is this latter sense of experience-present, understood as a subjective quality, that is most often found in the study of religion, insomuch as the outward behaviors and institutions are assumed merely to reflect an inner disposition that is beyond words. See **Phenomenology**.

Faith – [Latin *fides*, meaning trust, confidence, reliance] a term today commonly used alongside '**religion**', sometimes assumed to be the essential element to the religious life; sometime in fourteenth- and fifteenth-century Europe seems to be the first time we find 'faith' used as a synonym for 'religion'. In the **modern** sense, faith (as in **Wilfred Cantwell Smith**'s notion of 'faith in transcendence') is often juxtaposed to the social or institutional sense of religion (what W. C. Smith termed the 'accumulated tradition'), as in the distinction between 'spiritual' and 'religious' when the latter is assumed to denote the merely secondary, external, institutional or ritual elements whereas the former denotes what is assumed to be the personal and core element that is merely symbolized or manifested in the institution. Given the sixteenth-century Protestant reformers' efforts to criticize, and eventually to replace, the institutions and authority of Roman Catholicism, prioritizing faith over religious institution, and criticizing the latter for the manner in which it unnecessarily stifles the former, remains a common anti-Catholic, or pro-Protestant, form of argumentation.

Faith-Based Initiative – a program created early in President George W. Bush's first administration (2000–2004) that financially supports community service organizations that are run by local religious organizations. The relevance/controversy of this initiative is that, in the past, the US Federal government has avoided supporting religiously identified organizations that carry out social services (such as operating day cares, food banks, homeless shelters, programs for alcoholics, etc.) due to the understood separation of **Church and State** in the **First Amendment** of the US Constitution.

Family Resemblance – an approach to defining something, first described by the philosopher **Ludwig Wittgenstein**, that presupposes that no one characteristic is possessed by all members of a group but, instead, that a series of traits must be present, each to varying degrees, among the members of a group. See **Essentialism, Functionalism, Polythetic Definitions**.

First Amendment to the US Constitution – an amendment ratified in 1791 as a part of the Bill of Rights that prohibited Congress from interfering with the freedom of religion, speech, assembly or petition. 'Congress shall make no law respecting an establishment of religion, or prohibiting the free exercise thereof; or abridging the freedom of speech, or of the press; or the right of the people peaceably to assemble, and to petition the Government for a redress of grievances'.

Free Exercise Clause – portion of the **First Amendment** of the US Constitution that denies Congress the right to prohibit the 'free exercise' of religion. What exactly constitutes free exercise is unclear, however, and therefore open to debate. Congress does have the power to limit certain practices whether they are religious or not. A recent example of the debate is found in a 1992 US Supreme Court case over whether a city council in Florida could use their animal cruelty laws to curtail animal sacrifice as practiced by members of a Santeria group (Santeria, 'the way of the saints', is a Caribbean tradition that combines elements of African and Roman Catholic religious practices). See **Establishment Clause**.

Functionalism – the view that, rather than some internal quality, things are defined by what they do and can be studied in terms of the purposes that they serve or the needs that they fulfill. Functionalists can study the social, political or psychological role played by, for example, a **myth** or a **ritual**, examining how it functions either for the individual or how it contributes to maintaining an overall social structure into which the individual is placed. See **Essentialism, Family Resemblance**.

Greek – [*Graecus* meaning the name applied by the Romans to the people called by themselves Hellenes] the **Christian** text commonly known as the New Testament was written in a script known as common or *koiné* (pronounced 'coin-ay') Greek. It is important to note that words/concepts that were once prominent in the Hellenistic world of early Christianity, and therefore used in the production of these texts, eventually were translated into Latin, and then into the many languages that today comprise the text known as the Bible.

Hermeneutics – [Greek *hermeneutikos*, meaning translator or interpreter] the precise history of the term is unknown, though some trace it to the name of the Greek god Hermes (known by the Romans as Mercury) who served as a messenger for the gods; others trace it to Hermes Trismegistus, the Greek name for the ancient Egyptian god Thoth, said to have been the founder of alchemy and other such secret sciences. In any case, hermeneutics is that branch of study that deals with interpretation, both the act of interpretation as well as the academic study of the methods and **theories** of interpretation. Often associated with the interpretation of scripture, as in the long history of hermeneutics in the field of Biblical studies, hermeneutics presupposes that the object of study must be understood for its meaning and that this meaning can only be adequately understood if it is interpreted and translated in precise and correct ways. See **Phenomenology of Religion**, **Positivism** and **Reductionism**.

Hinduism – [Sanskrit *sindhu*, meaning river, especially the body of water known today as the Indus River (in northeastern India), hence the region of the Indus, which today also names the entire **nation-state** of India] the name given to the mass social movement found originally in the sub-continent that is today known as India and dates to up to 1,500 years prior to the turn of the era; those who practice Hinduism refer to it as **sanatana-dharma**; it is a term for indigenous Indian religions, and is characterized by a diverse array of belief systems, practices, institutions and texts. It is believed to have had its origin in the ancient Indo-Aryan Vedic culture, though this thesis is open to scholarly debate. Texts in Hinduism are separated into two categories: *shruti* (inspired [revealed scripture]) and *smriti* (remembered [epic literature]). The *Veda*, a body of texts recited by ritual specialists (brahmins) is considered *shruti*, whereas the *Bhagavad Gita* is considered to be *smriti*. Other *smriti* texts are the major epics: the *Ramayana* and the massive text known as the *Mahabharata*. Some of the commonly known deities are Vishnu, Brahma, Kali, Ganesha, Shiva and Krishna. Studies of Hinduism will often focus on the role played by the *dharma* system (social system of duties and obligations), the *caste* system (similar to a class system but inherited), beliefs in *karma* (social actions result in future reactions), *atman* (the name for one's soul or self) and *samsara* (the term for the almost limitless cosmic system of rebirths), and the central role of brahmins (a caste of **ritual** specialists).

History – [Latin *historia*, meaning narrative, account, tale, story] by 'history' we today mean at least two things: (1) a narrative (i.e., a tale with a beginning, middle and end) about the accumulated, chronological past that either demonstrates development over time or established lineage and (2) a more general usage that refers to the world of cause/effect in which unanticipated events intermingle with the intentions of **agents**. Saying that something is 'an element of the historical world' therefore implies that the present is the result of a series of past plans as well as accidents, which were themselves the results of yet other past plans and accidents. To say that something is 'historical' therefore means that it is **contingent**, i.e., depends on prior things happening and therefore could have been otherwise.

History of Religion(s) – although it may imply one studying the **historical** development of the **world's religions**, the term History of Religions most often names an academic discipline or approach thought by its practitioners to be distinct from other approaches to the study of religion (in North American it is often associated with scholars trained at the University of Chicago). In the German tradition, the field is known as ***Religionswissenschaft***, or the systematic, rational study (*Wissenschaft*) of religion, although turn-of-the-century generations of German scholars were part of what was then called the *Religiongeschichte schule* (the history of religions school of thought). This tradition was concerned with tracing the development of religion from its earliest stages to its modern form as well as turning methods previously used to study the scriptures of 'others' onto the texts of **Christianity** (known as the historical critical method, which approached the study of texts much like an archeologist might study artifacts: identifying and then sifting through their various compositional layers of a text, in an effort to understand the worlds of their various composers and their original audience). Although the North American field is indebted to these two traditions, no adequate English translation of *Religionswissenschaft* and *Religionsgeschichte* was apparent, hence the term Religious Studies was coined. For yet others, the term History of Religions was preferred, insomuch as it emphasized the historical (i.e., empirical, contingent) nature of the scholars of religion's data, in contradistinction from **theological** approaches to the same material. It is in perhaps this sense that the term is used in the name for the field's only international scholarly association, the International Association for the History of Religions (IAHR). For some, notably those influenced by the work of

Mircea Eliade, History of Religions came to designate a **herme-neutical** and **phenomenological** approach to the material, which gathered data by means of cross-cultural comparison yet assumed that all religious artifacts, symbols, behaviors and beliefs contained a common, deeply meaningful element that ought not to be merely **reduced** to their psychological or social elements. Perhaps for this reason one might place great significance on whether 'religion' appears in the singular (i.e., History of Religion, which names the study of a cross-culturally stable, analytic concept) or in the plural (History of Religions, which names a variety of different and thus empirically distinct instances). See **Comparative Religion**.

Human Nature – a concept – sometimes termed the human spirit, the human condition, the human heart, or the human **experience** – that asserts all human beings hold some **essential** characteristic(s) that is universal and thus not bound by any notions of time or space. All human beings, from the beginning of time and spanning the entire present world, are therefore said to share such characteristics, making these traits the defining element, or essence, of the human species as a group separable from all others. Some scholars of religion argue that religion, or religious experience, is the pre-eminent or fundamental aspect to this presumed human nature.

Human Sciences – those academic studies of minds, texts, social institutions, political organization and economic activity that seek to develop **theories** that explain human behavior rather than offer an interpretation of, or appreciation for, the meaning of the behavior or its various artifacts (such as texts, art, architecture, etc.). This classification of work carried out in the **modern** university provides an alternative to the traditional division of **social sciences** versus **humanities** insomuch as the human sciences groups together fields previously studied separately in either of these other divisions, understanding all elements of human social life to be subject to the tools of observation, analysis, generalization and explanation. Practiced as part of the human sciences, the study of religion seeks not to discover the meaning of religiosity but its causes and practical implications.

Humanities – an organizational title given to that area of the **modern** university that usually includes such academic disciplines as the study of literatures, languages, theater, philosophy, history – all of which are often presumed to study various expressions of the enduring human spirit as it is manifested in the conscious, intentional, and

most importantly meaningful, actions of **agents** in different historical periods and regions. Once taken out of **theological** studies, the
academic study of religion is often placed within humanities divisions of the university because its presumed object of study – religious experience – is often held to be a key ingredient to **human
nature**. See **Social Sciences**.

Idealism – a philosophical viewpoint that prioritizes mind or spirit over
matter or the physical world, the latter thought to derive from the
former; to be distinguished from **materialism**.

Ideology – first coined in late eighteenth-century France, the term stood
originally for the systematic study of ideas, or science of ideas, but
soon came to stand in for both a complete system of ideas, or what
we sometimes term a **worldview**, as well as an incorrect or false
system of ideas (the former a more descriptive use of the term
whereas the latter is a more normative use of the term). The term
obtained its best known and most critical usage in the work of **Karl
Marx**, where it was used to name the system of 'false consciousness' within which oppressed people labored. Today the term
retains this critical edge, though it is also used in a more neutral
fashion.

Induction – any form of reasoning in which a general conclusion is supported by the premises, but does not necessarily follow from them;
inductive logic begins with the observation of specific cases and
reasons to general conclusions based on this series of discrete observations. Classical scientific method, which prioritized observation and description, was thought to proceed inductively, in that a
general conclusion (for example, about the law of gravity) followed
from a series of experiments (such as repeatedly dropping an object
and observing its behavior). Inductive conclusions are only as sound
as the number of instances that support them (that is, how many
spotted dogs must one see before one is confident in concluding:
'All dogs are spotted'?), leading one to see that induction does not
provide certain, but instead probable, knowledge; distinguished from
deductive logic.

Inference – to derive a conclusion from something known or assumed to
be the case, knowledge which was itself gained by means of either
induction or **deduction**.

Intention – quality said to be possessed by **agents**; ability to have motivations, goals and desires that direct one's actions. Traditional literary

critics approached the study of texts in the effort to recover the original intentions of their authors, though a number of contemporary scholars question the direct linkage once generally assumed to exist between the meaning of a text and the intention of its author.

Inter-religious Dialogue – following the **colonial** age of **Christian** missionizing, in which the conversion of so-called 'heathens' was the goal, a more theologically and politically liberal movement began within Christianity in which some differences that came to be seen as merely secondary were put aside in favor of a search for more fundamental, **essential**, similarities among the **world's religions**. Mutual understanding, respect and appreciation therefore took over from a previous era's attempt to judge and convert. As practiced by some (particularly by **humanists**), the academic study of religion is seen as one component of the effort to identify and nurture such supposedly shared commonalities. For those who see the study of religion as part of the **human sciences**, such versions of the field are indistinguishable from liberal **theology**.

Islam – in Arabic meaning literally 'submission' (and one who submits is known as a Muslim), the name given to a collection of beliefs, practices and institutions that date to the sixth and seventh centuries **CE**, originating in the Arabian peninsula, which place importance on the role played by the Prophet Muhammad who is believed to have received, by means of recitations granted to him by an angel, the word of Allah (Arabic, 'the God') which is contained in their scripture, known as the Qur'an (sometimes written in English as 'Koran'). These revelations, which occurred in the area outside of the city of Mecca (today considered the central geographic site of Islam, toward which devout Muslims worldwide face when praying each day and to which they aim to make a pilgrimage at some point in their adult life), were eventually transcribed and today comprise the Qur'an's 114 suras, or chapter divisions, each of which have a number of verses. Merging indigenous Arabian cultural practices and beliefs with elements of **Jewish** and **Christian** belief, Muslims (those who submit to the will of Allah) understand Muhammad to have been the last in a long line of prophetic figures (stretching from Abraham to Jesus); he is understood to have been the 'seal of the prophets' (as in a stamp to close an envelope), all of whom conveyed the divine word, law and instructions of Allah. After establishing the first Muslim community in the nearby city of Medina, Islam spread successfully throughout much of what was then the known world, stretching across North Africa, Europe, and well into Asia. Today it can be found all throughout the world. Early on in its development, disagreements over such things as leadership succession led to a division, leaving two main sub-types: Sunni and Shi'ite (a third sub-type, Sufism, is considered the mystical aspect of Islam) – all of which have their own sub-types, often based on differing traditions of legal and textual interpretation.

Judaism – the name given to a collection of beliefs, practices and institutions that date at least to several hundred years prior to the turn of the era (though much further according to some members) and whose significant historical events transpired in the area of the world now known as the Middle East; although today considered a religious designation, to some it has always been merely an ethnic designation and – especially since the founding of the state of Israel in 1948 – for yet others it designates a national identity (sometimes designating all three at once). The terms 'Jew', 'Jewish' and 'Judaism' derive from the ancient Hebrew y'hudi which is itself a derivative of the proper name Y'hudah or Judah, which means 'celebrated' and was the name of the fourth son of one of the group's ancient patriarchs, Jacob, as well as the name for the familial line (that is, tribe) that is said to have descended from him. Although one might talk of ancient Hebrew religion (involving twelve ancestral tribes, a distinctive role for the members of a priestly tribe, the centrality of temple worship, the **ritual** of priestly animal sacrifice, a period of enslavement in ancient Egypt, and a belief in a divine mandate to settle 'the promised land'), after the Exilic period (in which it is held that, for much of the sixth century **BCE**, Hebrews were conquered by the ancient Babylonian empire [specifically, a group called the Chaldeans] and forcibly removed from their land) the centrality of textual interpretation, the role of the rabbi (Hebrew: master) and the place of the synagogue (**Greek**: assembly, as a translation for the late Hebrew, keneseth) came to supplant the prior place of the temple and priests. Along with legal traditions and traditions of rabbinic commentary, the main scripture is known as TANAKH, an acronym standing for the letters that signify the three main bodies of work that comprise what is sometimes called the Hebrew Bible: Torah (the Law, which comprises the first five books of the Hebrew Bible), Neviim (the writings attributed to the Prophets), and Ketuvim 'the writings' (such as the more poetic book of Psalms that is attributed to the patriarch and one-time Hebrew King, David). Today, Jews are found worldwide and the **modern** state of Israel (the so-called 'promised land') plays a particularly important role in the social identity for many Jews.

Li – Chinese term, associated with **Confucianism**, that names the rules of propriety (or proper form) associated with carrying out **ritual** and which influence all social interaction. See *Eusebia* and *Pietas*.

Linguistics – [Latin *lingua*, meaning tongue] the cross-cultural and comparative science of language as a human phenomenon, including phonetics, phonology, morphology, syntax, semantics, pragmatics and historical linguistics. In the late nineteenth century, when the cross-cultural study of languages was developing (once called Comparative Philology), it served as a model for the development of the early science of religion. Neither field was concerned about simply studying this or that language, this or that religion, but with studying language and religion as universal human phenomena (regardless which particular language one spoke or which specific religion one practiced), thereby necessitating the development of general **theories** of language and general theories of religion.

Materialism – a philosophical viewpoint that prioritizes matter or the physical world over mind or spirit, the latter being derived from the former; to be distinguished from **idealism**.

Metaphysics – [Greek *meta-*, meaning after, following + *physiké*, meaning nature and the world of production] the Greek phrase, *ta meta ta physika*, meant literally 'that which is after the physics', implying early cataloguing/placement of an untitled text by the ancient Greek philosopher Aristotle (384–322 **BCE**) after another of his works, entitled *Physics*, that dealt with the natural world or, what he termed, Being that is endowed with motion (that is, self-determination, motivation, or will [i.e., **agency**], something not possessed by art). The untitled text that followed *Physics* in early collections of Aristotle's works dealt with questions of origin and first principles; it traditionally went by the name of *Metaphysics*. What was therefore originally a term of sequential classification comes today to stand for that branch of philosophy that addresses questions of Being, reality, existence, the origins of the universe, etc. Today the prefix 'meta-' is often used to signify theoretical work, or work that examines assumptions that operate behind scholarship, as in the difference between developing a **theory** of religion as opposed to studying theories of religion themselves: meta-theory. See **Positivism**.

Modernity – although commonly used as a synonym for 'contemporary' or 'current', such terms as 'modern', 'modernism' and 'modernity' are used as technical terms to name a period in European, and later North American, history that could be said to develop from sometime in the seventeenth until the late twentieth century, characterized by certain intellectual, governmental, legal, economic, social, artistic and architectural movements. Given the different usages of the terms modern, modernism and modernity, a precise definition is difficult to provide; it generally denotes a period whose climax comes in the late nineteenth and early to mid-twentieth century, characterized by assumptions concerning how meaning and representation function as well as how social organization ought to take place. The rise of the scientific method, realist painting (based on perspective), the **correspondence theory of meaning,** industrialist and capitalist economies, and the **nation-state** all represent moments in the development of modernity. Contrary to **postmodernism**, understood as a mid- to late twentieth-century European movement associated with an emphasis on disjunction, difference, perspective, context, and the gap between a signifer (such as a red octagon with white trim and the letter S, T, O and P written inside it) and that which it is presumed to signify (come to a stop), modernism is associated with a confidence concerning the direct links between **intentions**, words and meanings.

Monothetic/Polythetic Definitions – deriving from **Greek** for either one, alone (mono-) or many, much (poly-) that are 'capable of placing', as in one-placement and many-placements. Monothetic definitions, which can be **essentialist** or **functionalist**, presume a limited set of necessary characteristics or purposes whereas polythetic (or what might also be termed multi-factoral) definitions identify a range of traits or functions, none of which is sufficient in order for the object to qualify as a member of a class. See **Family Resemblance**.

Muslim – see **Islam**.

Mysterium tremendum et fascinans – Latin phrase coined by the German Protestant theologian **Rudolf Otto** to name the awe-some (fascinating, full of awe) mystery that, he argued in his German work on comparative **theology**, *Das Heilige* (1917; translated as *The Idea of the Holy*, 1923), was the object common to all forms of religious **experience**.

Myth – [Greek *mythos*, meaning word, story or narrative] term whose current popular understanding can be traced back to an argument of Plato's in his ancient Greek dialogue entitled *The Republic*; 'myth' often designates fanciful, false or fictional narratives that are to be distinguished from historical narrative or rational discourses (Greek, *logos*). Sometimes used to refer to narratives that are transmitted orally and tell of supernatural beings that can accomplish deeds that humans cannot. For **idealist** scholars, myth, conceived as the expression of certain modes of thought, was understood to come before, and thus inspire, **ritual**. 'Myth' as a classification is now often used by **functionalist** scholars of religion to refer to any narrative that is used by a group of people to satisfy any basic need that a society or an individual may have.

Nation-state – [Latin *natio*, meaning stock or race, 'that which has been born'; as in native + Latin *status*, meaning position, the manner of standing, one's condition, as in the condition of a region or place) the name given to **modern**, large-scale social units that combine the earlier sense of an ethnic or ancestral group (a nation, clan or tribe) with the more recent political sense of a group organized around legal principles (such as those who possess citizenship not as a birthright but as an identity adopted by means of legal procedures). Often, nationalism, understood as an expression of one's political identity, is distinguished from patriotism with only the latter being understood as positive. This distinction is spurious for it is apparent that the same behavior (singing songs, marching, displaying flags and military hardware, engaging in nationalist rhetorics, presumptions that God is on one's side during a time of war, etc.) when practiced by one's enemies is classified as nationalistic whereas when practiced by one's own group of one's allies it constitutes benign patriotism.

Naturalistic Theories of Religion – as opposed to **theological** approaches to the study of religion that presume that the basis of religion is to be found in a supernatural source (such as God, the gods, etc.), naturalistic approaches presuppose that those beliefs, behaviors or institutions classified as '**religious**' are in fact mundane elements of the so-called natural world – that is, the **historical**, **cultural** world. In this sense, 'natural' does not necessarily carry the connotation of 'inevitable' or 'the way it ought to be' but, instead, is linked to an earlier sense of 'natural science' in that it is the systematic study of the empirical (observable with one of the five senses) world. Of course this is to be distinguished from what was once called 'natural religion' – the category used for those who infer the existence of God from the observation of the natural world, such as the so-called design argument (i.e., the complex workings of the natural world betray the existence of a design and a design necessitates the existence of a designer, much as the complex workings of a watch found on an isolated beach signifies the existence of a designer [or so first argued by the Christian theologian William Paley (1743–1805)]; contemporary Intelligent Design efforts to undermine **evolutionary** theory presume this very argument). Although early contributors to a naturalistic approach to religion can be dated to several centuries ago – notably the Scottish philosopher **David Hume**'s book, *A Natural History of Religion* (1757) – they began to flourish in the late nineteenth century and today involve the work of, among others, **psychologists, anthropologists, sociologists, political economists** and **cognitive scientists**.

Necessary – as used in philosophy, necessary is opposed to **contingent**; the former signifies something that is inevitable or required, by strictest definition, whereas the latter signifies something that may or may not be the case, depending on a variety of other, prior factors. For example, it is a biological necessity that human beings require oxygen in order to live; however, continuing to live is contingent upon breathing, eating, sleeping, etc. Sometimes necessary is distinguished from sufficient, as in a necessary cause versus a sufficient cause: the former signifies a prior factor that is required should certain results come about (such as the need for professors to publish their research in order to be awarded tenure), whereas the latter signifies a prior factor that alone will lead to the desired results (i.e., publishing research alone is not sufficient to be tenured, for teaching well is also required).

Orientalism – a term that has traditionally named a scholarly discipline, at its height in nineteenth-century Europe, that takes as its subject matter the study of the Arab world (the so-called Orient or what was once commonly known as 'Mystic East'), its history, language, and contemporary customs, religion and politics. It is this sense of the term that we today find in the name for the University of London's well-known School of Oriental and African Studies (SOAS), founded in 1916. More critically, 'Orientalism' now often stands for a particular attitude toward what scholars term 'the Other'; in this sense, most famously examined in a book by this title by the Columbia University literary critic Edward Said (1935–2003), Orientalism names a widespread strategy whereby groups create a sense of themselves as distinct from others by generating powerfully negative and easily reproduced caricatures and stereotypes of those from whom they see themselves to differ. In Said's analysis (*Orientalism* [1978]), the eighteenth- and nineteenth-century preoccupation among Europeans intent on studying 'the Orient' – learning its languages, mapping it, studying its **culture**, and writing novels about its mystery and danger – functioned to create a representation of the Arab 'Other' that reflected not actual traits in this part of the world, but, instead, functioned to reinforce a sense of superiority and order at home in Europe. Given that today the term 'Orient' no longer refers to the so-called Middle East (and such **modern** countries as Egypt, Israel, Syria, etc.) but to parts of Asia (e.g., China, Japan, Korea, etc.), it should be evident that this term is plastic and can be applied to whomever the apparent in-group sees as different from themselves and thus unknown.

Phenomenology – [Greek *phainomenon*, to appear] the descriptive and systematic study of that which appears or that which presents itself; to be distinguished from ontology [Greek *ontos*, being], the philosophical study of being or ultimate reality, as well as **metaphysics**. Although first developed in nineteenth- and twentieth-century European philosophy (notably the work of the German philosopher, Edmund Husserl [1859–1938]), early on scholars of religion adapted phenomenological methods to develop a technique for studying claims, symbols, practices and institutions that seemed to defy rational explanation (such as belief in an afterlife or rebirth). The term 'phenomenology of religion' is credited to the Dutch scholar, P. D. [Pierre Daniel] Chantepie de la Saussaye (1848–1920). This approach avoids assessing the truth or reality of such claims (because their truth is thought to reside in the subject's interior sentiments), studying instead what is assumed to be the public forms taken (i.e., that which appears to the observer's sense) by what is often termed a symbol's **essence** or a text's meaning. Phenomenologists of religion, many of whom would also be termed comparativists, therefore suspend judgment (i.e., are methodologically agnostic; see **Agnosticism**]) and work to describe what appears to them rather than judging it or criticizing it. They are therefore well-known for advocating empathy as well as the bracketing (or setting aside) of assumptions and preconceived notions when one confronts unfamiliar data. Phenomenolgocial method therefore presupposes both the objectivity of observers as well as their ability to identify with the experiences and meanings of the people they study. See **Experience, Hermeneutics, Positivism, Reductionism**.

Pietas – the Latin term, from which we derive the English word 'piety', that is commonly (although perhaps misleadingly) translated by many people today as 'religion'. In the classical world, 'piety', like ***eusebia*** and ***din***, denoted a quality that resulted when one fulfilled one's social obligations and duties, which involved everything from properly performing rituals toward the gods to treating one's superiors, peers and inferiors properly – that is, according to custom and the accepted rule of propriety (see *Li*). The common assumption today, prevalent in the Euro-North American world and especially within the **Christian** tradition, that 'religion' denotes an inner **faith** or **experience** therefore interiorizes or privatizes that which, in antiquity and in other cultures, was considered a public trait linked to observable behaviors one would or would not perform satisfactorily.

Political Economy – the systematic study (science) of the manner in which systems that govern power and privilege are interconnected with systems that govern patterns of exchange and the valuation of goods; the earlier name for what is today often referred to simply as economics.

Polythetic Definitions – see **Monothetic/Polythetic Definitions**.

Positivism – Although for some 'positivism' is used along with '**reductionistic**' to name a attitude toward the study of **culture** that – at least for those who fall in the **theological**, **hermeneutic** and **humanistic** traditions – is seen as overly reliant on the effort to reduce the meaning of a participant's testimony to observable, and thus predictable, facts, more properly it is termed 'Logical Positivism' – a term derived originally from the work of the early French social theorist, Auguste Comte (1798–1857), and which refers to an originally Austrian and German philosophical school of thought (which exerted great influence in North America as well) that dates to the early decades of the twentieth century. Members of the so-called 'Vienna Circle' of philosophers developed a system of rules for establishing which propositions were and were not meaningful and thus the proper topic of scientific discourse; their system thereby classified many of the traditional topics addressed within the field known as **metaphysics** (e.g., Does God Exist?) as meaningless. Their 'verifiablity principle' ensured that only those propositions that could conceivably be tested empirically or logically, and thereby found either to correspond to some observable state of affairs in the natural world or to obey the rules of logic (i.e., were verifiable), were meaningful – all others were classified as nonsensical or, as in the case of statements on morality, merely emotive. **Experience** and the use of human reason to organize experience and generalize from experience, were, therefore, the only basis for knowledge, and facts were understood to be independent of human consciousness and intention, thereby ensuring that objectivity was an attainable goal. Reaching the peak of their influence by the mid-twentieth century, it was realized that logical positivists' criterion of verifiability itself did not obey their own rules (the rule itself did not **correspond** to any empirically observable fact); this presented an empirical and logical problem that could not be overcome – although the philosopher of science, Karl Popper (1902–1994), revised this principle as the falsifiability criterion, whereby scientific propositions were those that could, conceivably, be empirically tested and, at least potentially, disproved. Until such a time as a proposition was disproved (such as 'All dogs have four legs'), it could be used 'as if' it was true, recognizing that one can never arrive at certain knowledge based on **induction**. Although few scholars of religion would today classify themselves as positivists in the earlier sense of the term, the goal of distinguishing participant claims from claims about participants – relying, to varying degrees, on the distinction between values and facts – yet remains for many scholars of religion.

Postmodernism – contrary to **modernism**, postmodernism denotes a skeptical attitude that follows on the heels of a confidence that was characteristic of pre-World War I Europe, the site of the rise of industrialization, the establishment of the **nation-state**, and capitalism. Influenced by **Existentialism**, postmodernism first takes hold in such areas as architecture and art, where disjunction is used to draw attention to the manufactured nature of all items of culture; because uniformity, such as windows on a building all being aligned with each other, is not natural or inevitable, but, instead, actively constructed, postmodernists draw attention to this constructive activity in a way previously unseen in art, architecture, etc. Contrary to realist art, then, postmodern works draw the viewer's eye to the composition of the artwork, aspects that had previously been overlooked. In literature and philosophy, postmodernism comes to name a movement that problematizes the previous attitude toward the inherent links between intentions, words and meanings; meaning-making therefore comes to be seen as an ongoing activity with direct relevance for context, such that postmodernists are often criticized as relativists.

Profane – [Latin *profanus*, from *pro* + *fanus*, meaning before, as in outside or in front of, the temple] considered the opposite of **sacred**;
that which was not admitted into the temple, or done while in the
temple, which extends to notions of not consecrated, **ritually** un-
clean, polluted or improper (as in 'profanity' used to signify
improper speech).

Prototype – The original or model on which something is based or formed;
something that serves to illustrate the typical qualities of a class or
group. For so-called Western scholars, **Christianity** is often uncon-
sciously the prototypical religion, providing the model by which one
judges other religions. Although **Islam** is not the worship of the
Prophet Mohammad, the name by which it was once known in
Europe – 'Mohammadism' – and even the name by which 'the
middle path' is commonly now known – '**Buddhism**' – provides a
useful illustration of how earlier European scholars used their knowledge of the centrality of Jesus Christ in something familiar to themselves, as the model for naming and thereby comprehending things
that were new and thus unknown. That prototypes are necessary
for cognition is not argued by contemporary scholars; instead, as
argued by the **anthropologist** Benson Saler, what is argued is that
prototypes are not to be understood or used as ideal cases. Rather,
they are working models that require adjustment when new information is acquired. The selection of features to be included in **family resemblance** definitions are generally thought to arise from a
prototype with which one happens to be familiar.

Psychology – the systematic study (science) of the mind or of mental
states and processes; psychology of religion is but one among a
number of subfields of the academic study of religion.

Q – [German *Quelle*, meaning source or origin] 'Q' is the shorthand
scholars use to name a document hypothesized to have existed and
been in distribution among members of the earliest groups that
eventually become known as **Christianity**. Based on the study of
the so-called synoptic Gospels (syn-optic = one eye, implying to
view together; a term given in the late eighteenth century to the
first three Gospels: Matthew, Mark and Luke) – which possess similar narrative structure and content, though which differ on significant details of the story of Jesus' life, teachings, and resurrection –
historians and literary theorists hypothesized that one or another of
the three existed first and that subsequent authors must have had
access to the earliest, from which they borrowed material. For those

who argue for the priority of the Gospel of Mark – and there are those who argue for the priority of Matthew – those passages that do not appear whatsoever in the text of Mark (which is by far the shortest Gospel) and which are common to Matthew and Luke (many of which are 'sayings' of Jesus, such as the Sermon on the Mount's Beatitudes ['Blessed are the poor...']) are thought to have derived from a separate document, also in circulation at the time but no longer in existence. Scholars who advocate this 'two source hypothesis', therefore, attempt, through their analysis of the texts in existence, to reconstruct Q.

Rational Choice Theory – a **modern** form of social theory applied to the study of religion, and derived from theories of economics that attempted to account for the means by which consumers made their selections among alternatives. Rational choice theory, favored by a number of US **sociologists** of religion, such as **Rodney Stark**, argues that such things as church membership are based on a series of sensible decisions made by participants, based on their assessment of costs and benefits (or gains and loses). For those benefits that cannot be had immediately (or in this life, such as justice), a series of compensators are drawn upon that make up for the lack of the primary goal.

Reductionism – an approach to the creation of new knowledge that attempts to account for one level of phenomena in terms of a more basic series of propositions, much as observations from the world of biology (such as monitoring the growth rate of cells) can be explained by reducing them to the language of chemistry, which in turn can be reduced to the theories of physics. In the study of religion, reductionism is often criticized for 'throwing the baby out with the bathwater'; in other words, those who presume that religion is *sui generis* argue that reducing religion to, for example, **sociology**, and thereby explaining it completely as a sociological phenomenon, misses the irreducibly religious character of the belief, act, symbol or institution. Although religion undoubtedly has a social dimension, as **Ninian Smart** would have argued, it cannot completely be reduced to sociology – or **psychology**, or **political economy**, for that matter. For yet others who consider religion to be a thoroughly human institution, there is no choice but to study it by means of reductionistic, **naturalistic theories** derived from such domains as psychology, sociology, etc. In fact, even scholars who favor non-reductionistic approaches have little choice but to reduce, since their cross-cultural work necessarily must use comparative categories, such as **Mircea Eliade**'s use of 'the **sacred**', by means of which the language of participants is reduced to the language of the analyst. See **Hermeneutics, Phenomenology, Positivism**.

Religion – the precise etymology (or historical derivation) of the **modern** word religion is unknown. There are, however, several possible roots from which the term derives. Most commonly, the ancient Latin words *religere* (to be careful, mindful) and *religare* (to bind together) are cited as possible precursors. Whereas the Roman writer Cicero (106–43 BCE) favored the first option, the later Christian writer Lactantius (250–325 CE) favored the latter. In his book, *The Meaning and End of Religion* (1962), **Wilfred Cantwell Smith**, who is among the more noted scholars to have investigated the category's history, suggests that both streams – one emphasizing the private disposition to be mindful whereas the other emphasizes the more objective sense of social processes that build identity – may have coalesced into the Latin *religio*. **Jonathan Z. Smith**, also among the scholars to have devoted attention to this problem, observes in an essay entitled 'Religion, Religions, Religious' (in Mark C. Taylor [ed.], *Critical Terms for Religious Studies* [1998]) that in Roman and early Christian Latin literature the nouns *religio* and *religiones*, as well as the adjective *religiosus* and the adverb *religios*, were all employed mainly with reference to, in his words, 'careful performance of ritual obligations' – as in the modern sense of, in his words, a 'conscientious repetitive action such as "She reads the morning newspaper religiously"'. If this is chosen as our origin for the modern term, then there is some irony in the fact that today it is often used to refer to an inner sentiment, affectation (e.g., religious **experience** and **faith**) rather than within the context of **ritual** (i.e., routinized behavior and participation in social institutions). As J. Z. Smith has pointed out, the fact that ethics and etiquette books immediately precede books on religion in the US Library of Congress catalog system may carry with it this earlier sense of religion as a form of carefully performed behavior. Regardless which etymology one chooses, the term 'religion' remains troublesome for those who presuppose some universal **essence** to lie beneath the term – whether that essence is, as W. C. Smith argued, 'faith in transcendence' (in distinction from the outer, 'cumulative tradition', as he phrased it) or whether it is some more specific item, such as famously argued by the Swiss Protestant theologian, Karl Barth (1886–1968), who criticized 'religion' (that is, what he understood as inessential outward ritual and institution) as sinful (inasmuch as it was

human beings trying to know God – whether those human beings
were or were not **Christian**), as opposed to the revelation of God
in Jesus Christ (which, he believed, was bestowed upon humans by
God). That this approach has little, if anything, in common with the
naturalistic, academic study of religion should be clear to the reader.

Religionswissenschaft – a German term that roughly translates as the
science of religion (*Wissenschaft* = systematic study of). See
Comparative Religion and **History of Religions**.

Ritual – a system of actions that, according to their practitioners, is used
by a group of people to interact with the cosmos and/or directly
relate to superhuman beings; these actions may consist of worship,
sacrifice, prayer, etc. Commonly understood as any set of actions
that is supposed to facilitate interaction between humans and superhuman beings. For **materialist** scholars, ritual is often presumed
to predate **myth** insomuch as routinized behaviors are thought to
provide the physical and cognitive conditions in which meaning systems (and hence mythic narratives) can take place. Scholars study
ritual behaviors in terms of their psychological, sociological, political, even their economic causes and implications. That some
behaviors one might classify as a 'habit' (for instance, regularly brush-
ing one's teeth) could just as easily be classified as a 'ritual' suggests
that there is a great deal at stake in how one classifies behaviors as
well as in the particular **theory** of behavior that one uses to guide
one's classifications.

Sacred – [Latin *sacer*, meaning set apart, dedicated, distinguished, as in
set apart from the public or mundane world]; to be distinguished
from **profane**. Although widely used as an adjective (e.g., sacred
texts) 'the Sacred' was a term of choice for **Mircea Eliade**, used to
describe that which is shared in common among all religions and
that which manifests itself in varied forms throughout the symbols
of the **world's religions**: the **experience** of the Sacred. Akin to
other **essentialists** who name the object of this experience as the
Holy (**Rudolf Otto**) or Power (**Gerardus van der Leeuw**), or even
religious experience (**William James**).

Sanatana-dharma – a compound **Sanskrit** term meaning the eternal or
cosmic system of duties (*dharma* = system of social obligations, as
in people 'doing their *dharma*'), implying a universal moral order
comprised of countless beings all diligently carrying out their proper
social and ritual action; it is the term used by some practitioners of

Hinduism to refer to their cultural practices as unchanging and divinely sanctioned.

Sanskrit – an ancient Indo-European language that began on the Indian subcontinent; somewhat like Latin once functioned in the Roman Catholic Church, it is today the **ritual** language used in the sacred texts of **Hinduism** and some of the texts of **Buddhism**.

Semiotics – the systematic study of signs and symbols as elements of communicative systems of behavior; a **theory** of how signs come to be meaningful, based upon **linguistic** theory which assumes that meaning is not an **essential** quality expressed by symbols but, instead, the result of relationships established and managed by means of structures (such as a grammar or the rules of a game). See **Structuralism**.

Social Sciences – an organizational title given to that area of the **modern** university that usually includes such academic disciplines as economics, **sociology, psychology, anthropology** and political science; in distinction from the **humanities**, the social sciences seeks to generate testable **theories**, often on a model similar to the way knowledge is gained in the so-called 'hard sciences' (e.g., chemistry, physics, etc.). Since the social sciences study the actions and motivations of conscious subjects, they are sometimes known as the 'soft sciences', since their findings are sometimes critiqued as more interpretive and thus open to debate than the work carried out in other sciences. Because many in the academic study of religion understand their object of study to refer to an inner **experience** of tremendous meaning to the participant, this field is most often placed within the humanities, though it is occasionally found in social sciences divisions of the university. See **Human Sciences**.

Sociology – [Latin *socius*, meaning companion + *logos*, meaning word, speech, discourse, reason] the science or the study of the origin, development, organization and functioning of human society; the science of the fundamental laws of social relations and institutions. The sociology of religion is but one subfield of the academic study of religion.

Structuralism – developing from out of the context of the structural **functionalist** approach, once dominant in cultural **anthropology**, structuralism names an approach to the study of meaning systems (such as language or **culture**) much associated with the ground-breaking work of the French anthropologist Claude Lévi-Strauss (b. 1908).

Structural functionalists used a biological metaphor to understand social systems, seeing them as comprised of discrete and empirically observable units, such as: kinship systems, systems that govern the exchange of goods, system to govern the payment of debts, and **rituals** (such as rites of passage from one status to another) – all of which could be observed by the ethnographer (one who writes about another group of people). In playing their various roles and each fulfilling their separate function, they collectively contributed to the overall well-being of the unit. Contrary to this position, structuralists argue that the structure is not in the external world but, instead, is in the human mind itself. Through studying paired oppositions that occur in such things as rituals or myths (such as up/down, in/out, male/female, light/dark, cooked/raw, sacred/profane, etc.), Lévi-Strauss argued that scholars could decode the means by which groups of people set about ordering their worlds, classifying its components, establishing their relationships, and in the process making the world sensible and inhabitable. In this regard, structuralism owes much to the Swiss scholar, Ferdinand de Saussure (1857–1913), whose early work in **linguistics** and **semiotics** provided the basis for a critique of the **correspondence theory of meaning**. Saussure argued that, for example, the symbol 'i' means what it does because it is placed in relation to the symbols 'h' and 'j' (when understood in terms of this thing we know as 'the alphabet') or in relation to, say, the symbols 'f' and 't', as in the thing we know as the word 'fit'. And 'fit' has meaning because it is in relation to 'fat'; in other words, meaning is not an **essential** trait (that is, 'i' does not correspond to some i-ness); instead, meaning is the result of a series of relations of similarity and difference that is established by an overall structure which is itself **contingent**.

Sui generis – [from Latin, designates a thing that belongs to its own kind; peculiar; unique; self-caused] this term has been used to designate the claim that religion or religious **experience** is of a kind wholly unique and thus irreducible. If religion is *sui generis* then it is a thing of a kind incomparable with any other social institution or practice and therefore cannot be explained using a **naturalistic theory of religion**. Arguments for the unique nature of religion were successfully used in the 1950s and 1960s to help establish autonomous Departments of Religious Studies – insomuch as the studies of **anthropologists** or **sociologists**, to name but two, were thought to overlook and obscure the irreducible element (or **essence**): religious experience. See **Reductionism**.

Theism – [Greek *theos*, meaning god] a philosophical position to name a family of belief systems that presuppose the existence of God or gods; sub-types could include such belief systems as monotheism (belief in one God) or polytheism (belief in many gods); to be distinguished from **agnosticism** and **atheism**.

Theodicy – [Greek *theos* + *diké*, meaning the justice of god] term coined by the philosopher Gottfried Wilhelm von Leibniz (1646–1716) to name the problem of justifying belief in the goodness of an all-powerful divine being in light of routine empirical observations of what could be called evil in the world. Traditionally, the problem has been divided by European philosophers into three propositions that, it is argued, cannot all be held simultaneously (what is called a trilemma, as opposed to a mere dilemma): (1) God is all-powerful (omnipotent); (2) God is all-loving (omnibenificient); (3) Evil exists. Solving the problem of evil therefore requires one either to forsake one proposition in favor of the other two (e.g., evil exists because God is not all-powerful) or to adapt one or more propositions such that the trilemma is avoided (e.g., although it seems like evil to us, to God it is not). **Theistic** philosophers have also worked to develop ways to distinguish among types of evil that need to be addressed, such as moral evils, for which an intentional **agent** can be held accountable (and the evil thus explained as the result of free will) and natural evils, such as earthquakes (which do not appear to be the result of an intentional agent's actions). Although this is largely a **Christian** philosophical issue, **Max Weber** argued that the **Hindu** doctrine of karma (the cosmic law of actions and reactions, in which past deeds are thought to influence one's future rebirth status) was but one more attempt to address the problem of evil.

Theology – [Greek *theos*, meaning god + *logos*, meaning word, speech, **discourse**, reason] taken from the Greek, this term designates the academic discussion and study of God or the gods; 'theology' is commonly used today to signify the systematic study of **Christian** dogmas and doctrines, as carried out by a member of the group, but can be applied to any articulate and systematic discourse by members of a particular religion concerning their own tradition's meaning or proper practice or their tradition's view of others. It is to be distinguished from an **anthropological** approach to the study of religion in which human behaviors, not the actions of the gods, are the object of study.

Theory – [Greek *theoria*, meaning to look at, implying to observe, to
    consider, to speculate upon] a term that presupposes a distinction
    between reflection upon principles and causes as opposed to a
    form of practice; sometimes used as synonymous with philosophy,
    viewpoint or speculation, it can, however, be defined in a techni-
    cal, scientific manner to signify a series of logically related and test-
    able propositions that aim to account for a certain state of affairs in
    the observable world. Meta-theory (see **Metaphysics**) generally sig-
    nifies rational reflection upon the principles that underlie theoreti-
    cal work. For Marxist scholars (some of whom are members of a
    school of thought known as critical theory), the apparent separation
    between theory and practice is problematic, for they hold that theory
    too is a form of practical labor, and theory relies on practice which
    is itself directed by theory; they therefore often employ the term
    'praxis' to signify the correlation of, and dialectical relationship be-
    tween, these two seemingly distinct domains.

Worldview – term popular among some scholars as the wider category of
    which religion is but one instance; contrary to those who see reli-
    gion as **sui generis**, 'worldview' is a term used by some of those
    who understand religion to be but one among many means whereby
    human beings construct a coherent environment in which to carry
    out a meaningful existence by drawing/building on a series of com-
    mon assumptions, tales, behaviors and institutions – all of which
    enable them to organize and thereby experience their world as
    having order, sense, purpose, direction, etc. According to **Clifford
    Geertz**, in his well-known article, 'Religion as a Cultural System', a
    worldview is 'the picture [a group has] ... of the way things in sheer
    actuality are, their most comprehensive ideas of order'. **Ninian Smart**
    is among the best examples of a scholar of religion who argued that
    religion, along with such other systems as Marxism, constitutes a
    worldview. See **Ideology**.

World Religions – although taken for granted today, this term came to
    prominence only in the late nineteenth and early twentieth centu-
    ries and is today used to organize information for what are arguably
    the most popular courses and classroom resources in the study of
    religion. However, as made evident by such scholars as **Jonathan Z.
    Smith** and **Tomoko Masuzawa**, prior to the rise of this term, and
    the assumption that a diverse collection of beliefs, behaviors and
    institutions across the globe share a specific number of similarities,
    making them all members of the same family, one might expect to
    find Europeans using the term 'world religion' (note the use of the

singular), referring to Christianity as 'the true religion' that spans the world – a designation that implicitly contained a theological judgment concerning its superiority. Earlier designations grouped the information in terms of: ours and theirs, such as 'we' being Christian and 'they' being 'heathens' (those outside the city, who inhabit the heaths, i.e., rural areas); religions were also distinguished based on those that were revealed (those claiming divine revelation as their source) as opposed to those that were **natural** (in which people **inferred** the existence of god[s]); later, there were those that were **national**, as in those that were limited to a specific ethnic group, as opposed to those that successfully spread to other regions (making them, as noted above, a world religion). Once the plural term 'world religions' arises, the number of traditions included within the family starts out rather small but steadily grows over time, such that today one can easily find a fairly long list of world religions, which includes **Hinduism, Buddhism, Confucianism**, etc. The movement for **inter-religious dialogue** is based on the assumption of cross-cultural similarities among members of this family.

Scholars

As with the technical terms, the first time in each chapter that a relevant scholar's work is discussed that person's name is printed in bold text, signaling that a further discussion of this person's work, along with sample quotations, can be found in this section of the book. Also, just as in the chapters, technical terms used in this section are also bolded, so that readers can find definitions of these terms in the previous section of the book.

William E. Arnal

Trained as a scholar of **Christian** origins and specializing in the study of 'Q' – which stands for the German word for source or origin, *Quelle*, used to name a source document comprised of sayings of Jesus that scholars theorize must have existed in the earliest years of the social movement that comes to be known as Christianity – William Arnal's interest in Marxist social theory has led him to write considerably further afield than many scholars who work on early Christianity. Arnal carried out his doctoral work at the University of Toronto, under the direction of John Kloppenborg, the internationally noted Q specialist, earning his PhD in 1997, with his dissertation winning the Governor General's Gold Medal. He is widely published in the field's leading periodicals and has held academic appointments at New York University, the University of Manitoba, and is currently Associate Professor of Religious Studies at the University of Regina in Saskatchewan, Canada. Arnal has served as Vice-President of the North American Association for the Study of Religion (NAASR) and as the English-language editor for Canada's primary academic journal in the field, *Studies in Religion*.

Arnal on Religion

Unlike those who classify themselves as New Testament scholars, Arnal is among a group of scholars who study the texts and the context of early Christianity in order to shed light not on the meaning these texts might have had for their early writers or for their modern readers but, rather, on the Hellenistic social world from out of which Christianity began. Hence, as a scholar of Christian origins, Arnal employs social **theory** to help account for the shape taken by the early community and its spread. Although trained in the traditional tools of languages and textual criticism, it is his interest in theory that has led Arnal to work more broadly in the study of religion; his interest in the political implications of the category 'religion' itself prompts one to think of him as a meta-theorist. While being among the leaders of a new generation of specialists in early Christianity, Arnal has also developed a readership among those who are not necessarily specialists in Christian origins. More than likely it is the enduring **theological** presumption of the privileged place occupied by Christianity in the history of the world that prompts few scholars of early Christianity to consider themselves to be contributing to the wider field of religious studies. In fact, in many university curricula the designation '**History of Religions**' (or '**world religions**') is applied only to those who study religions outside of the so-called Western Religions – those 'others'

not identified with either **Judaism** (prominent for its role in the beginnings of early Christianity) and Christianity. We see here the remnants of
the Christian (generally Protestant) seminary model on which the academic study of religion was originally founded. Countering this long-standing
theological trend, Arnal's interest in social theory and critique guarantees
that, despite working on a specific data domain (that is, the texts of early
Christianity), his work constitutes an application of more general theories
regarding how groups contest identity and resources. A suitable example
is his latest book, *The Symbolic Jesus* (2005), which applies political and
discourse analysis to modern scholarly representations of Jesus' Jewish
identity, in an effort not to recover the authentic Jesus (as generally carried out in research on the historical Jesus) but, instead, to investigate
one instance of how symbolic representations are employed as a strategy
whereby contesting groups reproduce themselves.

'[O]ur definitions of religion, especially insofar as they assume a privatized and
cognitive character behind religion (as in religious belief), simply reflect (and
assume as normative) the West's distinctive historical feature of the secularized
state. Religion, precisely, is not social, not coercive, is individual, is belief-oriented
and so on, because in our day and age there are certain apparently free-standing
cultural institutions, such as the Church, which are excluded from the political
state. Thus, **Asad** notes, it is no coincidence that it is the period after the "Wars
of Religion" in the seventeenth century that saw the first universalist definitions
of religion; and those definitions of "Natural Religion", of course, stressed the
propositional – as opposed to political or institutional – character of religion as a
function of their **historical** context... The concept of religion is a way of demarcating a certain socio-political reality that is only problematized with the advent
of **modernity** in which the state at least claims to eschew culture per se.'
— 'Definition', in *Guide to the Study of Religion* (2000)

'[T]here is no such thing as religion in the world. Of course this may be said of
any taxon, but in the case of "religion", the formulation of the category has more
to do with the normative interests of modernity than with the intellectual or
theoretical motives of students of religion. "Religion" is an artificial agglomera-
tion of specific social behaviors, whose basis of distinction from other social
behaviors is a function of the specific characteristics of modernity.'
— 'The Segregation of Social Desire: "Religion" and Disney World',
in *Journal of the American Academy of Religion* (2001)

Talal Asad

Talal Asad, a post-colonial theorist and **anthropologist**, is among a generation of scholars deeply influenced by – and who has significantly furthered – the work of such scholars as the French intellectual, Michel Foucault (1926–1984), and the American (though born in Jerusalem and raised in Cairo, Egypt and Palestine) scholar of comparative literature, Edward Said (1935–2003). Moreover, he is part of a recent trend in anthropology – best exemplified in the work of James Clifford – in which the object of focus has turned from the so-called native to the means by which the ethnographer (one who writes [Greek: *graphé*] about another group of people [Greek: *ethnos*]) comes to know the native – that is, the ethnographer's tools, questions, categories, assumptions, etc. Like Foucault, Said – whose groundbreaking book *Orientalism* (1978) was among the first to introduce some of Foucault's early work to the North American readership – was interested in the intersection between systems of knowledge (such as classification systems) and systems of control (as in ways of asserting political power and influence). Thus, Foucault's thoughts concerning the complex inter-relations between knowledge/power were worked out by Said with regard to the manner in which early **modern** Europeans developed a way of understanding themselves and their worlds in relation to what they understood themselves not to be – defined in relation to politically useful stereotypes about the so-called '**Orient**', the name once applied by Europeans to the world of Arab language and culture. Working in this tradition, Asad is an essayist whose work explores the ways in which systems of knowledge and systems of discipline interact to produce specific ways of talking about, and thereby organizing, the world. Of the many classifications used by our own **culture** to enable us to know something about the world in which we live, Asad is perhaps best known for his focus upon the distinction between the **sacred** and the secular and the manner in which this distinction helps to make possible a specific sort of social identity: the modern **nation-state**.

Asad on Religion

Like **Tomoko Masuzawa**, Asad could be considered a meta-theorist of religion – or, better put, an archaeologist or anthropologist of the category '**religion**'. His interests revolve not so much around studying the behaviors and institutions of religious people 'in the field', as an anthropologist of religion might once have done, but, instead, they involve studying the manner in which the modern, European distinction between what is considered the sacred and the secular helps to make certain sorts of social and political worlds possible and thereby knowable – in other words, how generic things in the world become items of significance and thus of **discourse**. As such, Asad is among a relatively small group of scholars currently working on the **history** and implications of the classification 'religion' itself. His work therefore does not endeavor to arrive at a more adequate definition of religion conceived as a universal element of human minds or cultures; rather, it inquires into the implications of dividing the world up in just that way, examining what sort of social life is made possible by the apparently intuitive ability some people in the world think they have for distinguishing a religious from a political event, or a private from a public action. Tracing the history of the category religion to the European world, his work therefore also involves paying close attention to the manner in which this category is developed and exported, through **colonialism**, along with other aspects of Euro-North American culture and how it is applied to others in the world who do not necessarily employ it in their acts of self-classification. Although Asad does not advocate a pre-linguistic moment when reality in itself could once have been experienced – prior to limiting classifications systems, much as someone like **Rudolf Otto** might have argued – he nonetheless draws attention to the practical work invariably carried out by all knowledge/power systems.

> 'In what follows I want to examine the ways in which the theoretical search for an **essence** of religion invites us to separate it conceptually form the domain of power. I shall do this by exploring a universalist definition of religion offered by an eminent anthropologist**: Clifford Geertz's** "Religion as a Cultural System". My **intention** ... is to try to identify some of the historical shifts that have produced our concept of religion as the concept of a transhistorical essence – and Geertz's article is merely my starting point... My argument is that there cannot be a universal definition of religion, not only because its constituent elements and relationships are historical specific, but because that definition is itself the historical product of discursive processes.'
>
> — *Genealogies of Religion: Discipline and Reasons of Power in Christianity and Islam* (1993)

> 'What interests me in particular is the attempt to construct categories of the secular and the religious in terms of which modern living is required to take place, and nonmodern peoples are invited to assess their adequacy. For representations of "the secular" and "the religious" in modern and modernizing states mediate people's identities, help shape their sensibilities, and guarantee their **experiences**.'
>
> — *Formations of the Secular: Christianity, Islam, Modernity* (2003)

Pascal Boyer

Originally trained in Paris and then at Cambridge University, Pascal Boyer currently teaches in both the Departments of Psychology and Anthropology at Washington University in St. Louis, Missouri. While his earlier work was carried out in cultural **anthropology**, his main area of interest is how human memory works – how ideas are acquired, stored and transmitted – both within individuals as well as in those collections of individuals that we know as **cultures**. As with many who today are part of the cognitive science of religion (a relatively young field but one that has produced a surprising amount of research over the past decade), his work, which is based on the assumption that human minds have **evolved** over time to function as they currently do, combines traditional anthropological fieldwork with laboratory experiments, in an effort to develop explanatory **theories** of religion's origins and function, as opposed to interpretations of its meaning (as in **hermeneutical** studies). Although hardly the only **naturalistic** way to explain the causes and functions of religion, those scholars of religion who are today grouped together under the banner of **cognitive** theory have certainly proved to be among the most organized and ambitious of those working toward a theory of religion. Drawing upon findings from recent cognitive psychology (notably such fields as early childhood studies), **linguistics**, and theories of mind, a loosely knit group of scholars of religion, philosophers, psychologists and anthropologists – working both in Europe and North America – have, since the early 1990s, rather quickly developed a coherent, collaborative research project. Although once primarily associated with the ground-breaking, co-written work of E. Thomas Lawson and Robert N. McCauley (such as their *Rethinking Religion: Connecting Cognition to Culture* [1990] or their more recent *Bringing Ritual to Mind: Psychological Foundations of Cultural Forms* [2002]), today a variety of new, and largely young, scholars are now associated with what has become one of the most active and intellectually rigorous subfields within the modern study of religion. Currently, the model offered by Boyer has had the most impact among cognitivists (though the fieldwork-based theory of the Irish anthropologist, Harvey Whitehouse, has grown increasingly influential as well).

Boyer on Religion

Boyer argues that human beings' minds are wired in such a way that a slightly counter-intuitive idea (about, say, what an **agent** can and cannot do) is particularly appealing to human memory systems. Deviate too far from our evolutionarily derived common sense, or intuitive, expectations concerning such things as agency, and such novel ideas cannot compete very well in what is a pretty competitive economy of ideas and sensations circulating in the brain (that is, they are easily forgotten and thus not retained, much less transmitted). As Boyer argues in such books as *The Naturalness of Religious Ideas* (1994) and *Religion Explained* (2001), beliefs in the existence of beings who are very similar to how human beings see themselves (as in their appearance, the extent to which they can act, etc.) but who, for instance, do not die, know everything, and are not limited by the usual constraints of a body, are very appealing to our memory systems, thus making ideas about such beings easily remembered, which gives such ideas a considerable competitive advantage over other ideas when it comes to their transmission from one mind to another. To borrow from another anthropologist, Dan Sperber, who uses an epidemiological metaphor (as used in *Explaining Culture* [1996]), Boyer theorizes that these minimally counter-intuitive ideas stand out just enough to make them catchy. The underlying assumption, here, of course, is that, at the end of the day, those things that we study when we examine culture can be **reduced** to sets of ideas retained in, and transmitted between, human minds – ideas such as whether it is worthwhile to memorize and recite a particular set of texts, let alone memorize and retain the specificity of the character set in which it is encoded. Therefore, among the most basic ways of accounting for cross-cultural similarities, including those beliefs and practices we commonly call **religion**, is to account for the origination and transmission of these ideas, especially those ideas that fit the minimal requirements of being modestly counter-intuitive.

'Why do people have religious ideas at all? And why *those* religious ideas? These should be crucial questions for cultural anthropology; indeed, they are among the questions an uninformed outsider would assume are central to anthropological inquiry. As it happens, the problem is generally ignored in the discipline, though this neglect is a relatively recent phenomenon. The founders of **modern** anthropology had precise explanations for the appearance of religious notions. These hypotheses, however unsatisfactory, were at least a springboard for more refined speculations. Modern anthropology, by and large, is much less daring in its approach to religious representations.'
     — *The Naturalness of Religious Ideas: A Cognitive Theory of Religion* (1994)

'It is ... a hallmark of the "modern mind" – the mind that we have had for millennia – that we entertain plans, conjectures, speculate on the possible as well as the actual. Among the millions of messages exchanged, some are attention-grabbing because they violate intuitions about objects and beings in our environment. These counter-intuitive descriptions have a certain staying power, as memory experiments suggest. They certainly provide the stuff that good stories are made of. They may mention islands that float adrift or mountains that

digest food or animals that talk. These are generally taken as fiction though the boundary between a fictional story and an account of personal **experience** is often difficult to trace. Some of these themes are particularly salient because they are about agents. This opens up a rich domain of possible inferences.'
— *Religion: Explained: The Evolutionary Origins of Religious Thought* (2001)

Wendy Doniger

Originally trained as a dancer, Wendy Doniger completed her undergraduate work at Radcliffe College in 1962 and completed two doctoral degrees, at Harvard and Oxford Universities, specializing in **Sanskrit** and Indian studies (what is sometimes called **Oriental** studies). She has held teaching positions at Harvard, Oxford, the University of London, and UC Berkeley and has taught at the University of Chicago since 1978, where she is the **Mircea Eliade** Distinguished Service Professor of the **History of Religions**, in the Divinity School, as well as holding appointments in the Department of South Asian Languages and Civilizations as well as the Committee of Social Thought. In 1985 she was the president of the American Academy of Religion (AAR), the field's primary professional association in North America. She has written extensively (publishing earlier in her career under the surname O'Flaherty) on the religions of India, in particular the study of ancient **Hindu** myths but she has also translated into English a number of key ancient Hindu texts (including the *Rig Veda* as well as the *Kamasutra*) along with modern works of scholarship (such as the multi-volume French work of Yves Bonnefoy, *Mythologies* [1981]). Her interest in cross-cultural, comparative work extends well beyond the myths of India; her general interest in such topics as gender, sexuality, and personal/social identity enables her to do comparative work in a wide range of historical periods and cultural settings, evident especially in her later works.

Doniger on Religion

Although as a comparativist the themes of difference and similarity (what could also be termed the particular and the universal, the strange and the familiar, or the far and the near, as in **Clifford Geertz**'s sense of experience-distant and experience-near; see **emic/etic**) circulate through much of Doniger's work, difference tends eventually to take a back seat in her effort to illuminate a deeper degree of similarity that persists despite differences in context and content. For example, in **myths** – a particular type of narrative that she holds to function similarly to a number of other narrative types, such as legend, folklore, etc. – she finds specifically religious and thus universal questions of deeply held belief and human meaning raised, suggesting a uniformity beneath myths' variable contents and different themes. She presumes that there are levels of human **experience** that come before such things as the changing conventions of language, **culture** and **history** – making such pre-linguistic **experiences**, she argues, the basis upon which our shared **human nature** is based.

Scholarship on myths that avoids **reducing** them to something other than the expression of such pre-cultural meanings (that is, scholarship such as **Freud**'s reduction of myth to a social mechanism used by groups to vent, in a harmless manner, socially dangerous anxieties), can therefore recover, and in the process nurture, this shared humanity as it is expressed in the religious imagination of humankind (somewhat akin to Mircea Eliade's notion of a 'creative **hermeneutics**' that, as he argued, could lead to a new humanism). As such, Doniger's body of work provides an excellent example of a **humanistic** approach to the study of religion – one that invests much energy in describing the specificity of, and thus differences among, the various objects under study, attempts to avoid reproducing a specific **theological** viewpoint, yet one that nonetheless attempts to recover their shared, deeper meaning (rather than simply their social, psychological, or economic function).

> '[A] myth is not so much a true story as a story on which truth is based, a story which people may infuse with their truth.'
> — *Other People's Myths: The Cave of Echoes* (1988)

> 'It is customary in scholarly approaches to myth to begin with a definition. I have always resisted this, for I am less interested in dictating what a myth is (more precisely, what it is not, for definitions are usually exclusivist) than in exploring what myth does (and in trying to demonstrate as inclusive a range of functions as possible). Defining myth requires building up the sorts of boundaries and barriers that I have always avoided... The key game of cross-cultural comparison lies in selecting the sorts of questions that might transcend any particular culture. Some people think that there are no such questions, but some think, as I do, that worthwhile cross-cultural questions can be asked.'
> — *The Implied Spider: Politics and Theology in Myth* (1998)

Mary Douglas (1921–2007)

Born Margaret Mary Tew in Italy (while her British parents were on their way back to Burma, where her father worked in the Indian Civil Service for the British government), Mary Douglas was one of the twentieth century's most influential **anthropologists** and scholars of classification systems and institutions. She obtained her PhD from Oxford University in 1951, has carried out fieldwork in, among other places, the Congo, and has held teaching positions in both the UK and the US. Although she is also known for her work on how institutions function, and, in more recent years, she turned her attention to studying Biblical texts as literature, Douglas is perhaps still best known for her influential 1966 book, *Purity and Danger*, which was a cross-cultural study of **ritual** systems of cleanliness, pollution and taboo (a term that entered English in the late eighteenth century, as a result of Captain James Cook's travels in the Polynesian islands, meaning 'specially marked', as in set apart, forbidden, or even consecrated).

Douglas on Religion

Assuming that systems of purity or cleanliness, rather than being primarily concerned with establishing hygienic conditions, functioned instead to put into practice a system of order on an otherwise non-ordered world, Douglas studied systems of allowable and disallowable behaviors – such as the famous dietary codes as found in the Hebrew Bible's books of Deuteronomy and Leviticus. Her conclusions, well in line with developments at this time in such other fields as **Linguistics** and **Semiotics**, concerned the manner in which relationships articulated by means of human symbol systems helped to establish meaningful conditions in which human life could take place. Religion, for Douglas, was therefore the name given to but one collection of beliefs, behaviors and institutions that helped to orient and regulate social life, ensuring that certain behaviors could be understood as meaningful, memorable, and thus repeatable. Her early work on dietary codes was therefore but one occasion to demonstrate how human communities actively constitute their environments. It also provided an opportunity to examine what happens when human map-making inevitably fails, for, as a human system that reflects **contingent** preferences, this map-making activity is presumed never to be entirely adequate. Thus, reality continually presents cases unanticipated in our classification systems, puzzling cases that cannot but be understood as anomalies. This in turn ensures that all well-functioning classification systems require such categories as taboo (reserved for objects or actions that defy or conflate the usual categories, such as a bird with feathers that cannot fly). It is for this reason that all classification systems are understood as provisional and tactical, for in their application they are continually being reinvented and thus under construction.

'If we can abstract pathogenicity and hygiene from our notion of dirt, we are left with the old definition of dirt as matter out of place. This is a very suggestive approach. It implies two conditions: a set of ordered relations and a contravention of that order. Dirt then, is never a unique, isolated event. Where there is dirt there is a system. Dirt is the by-product of a systematic ordering and classification of matter, in so far as ordering involves rejecting inappropriate elements.'

'Defilement is never an isolated event. It cannot occur except in view of a systematic ordering of ideas. Hence any piecemeal interpretation of the pollution rules of another culture is bound to fail. For the only way in which pollution ideas make sense is in reference to a total structure of thought whose key-stone, boundaries, margins, and internal lines are held in relation by rituals of separation.'

— *Purity and Danger: An Analysis of the Concepts of Pollution and Taboo* (1966)

Emile Durkheim (1858–1917)

There may be no more influential figure in the study of religion than the late nineteenth-century French scholar, Emile Durkheim – considered to be one of the founders of the modern academic discipline of **sociology**. Although not all scholars today study religion sociologically, along with the **political economist Karl Marx** and the social **psychologist Sigmund Freud**, Durkheim is certainly among a very small group of writers who have had a tremendous impact on the modern field. Prior to scholars such as Durkheim, the now-taken-for-granted role that society plays in shaping individual consciousness and behavior was not so apparent to scholars. For this reason, his 1897 sociological study of the causes of European suicide helped considerably to legitimize sociology as a science. In that work, Durkheim argued that, unlike previous studies that argued that suicide resulted from individual decision or malady, the suicide rate was inversely correlated to the cohesiveness of a person's social group; that is, the higher rates of suicide among Protestants, as opposed to Roman Catholics and **Jews**, could be explained as a result of the former group's emphasis on the lone individual as opposed to the greater sense of social unity evident in the latter two (of which Jews were, for Durkheim, the strongest example since their communities in Europe were, historically speaking, set apart and, of strict necessity, much more self-reliant and cohesive). This leads to a crucial sociological insight contributed by Durkheim, one that is still provocative of thought: **religion**, he concluded, functioned as a 'prophylactic' against suicide not because of what it does or does not preach or teach to its adherents (in other words, not because of its content) but, instead, because of the role its all-consuming **rituals** and institutions play in bringing individuals together as a group, thereby providing them with not only a sense of belonging but also a sense of what it is to be a particular sort of individual.

Durkheim on Religion

Durkheim's explanatory **theory** of religion, to be distinguished from an interpretive approach that investigates what religion means, provides an excellent example of scholarship that **reduces theological** claims to science – in his case, to the language of sociology. In Durkheim's analysis, the rituals and institutions of religion are fundamental sites where social groups are formed; in the midst of the common behaviors (rituals) and heightened emotions characteristic of large social groups (what Durkheim termed 'collective effervescence', such as we find today at large sporting events, the so-called crowd phenomenon that can be found today anywhere from family celebrations to sporting events and nationalistic celebrations), the individual directly **experiences** the group and him/herself as a member. It is at such times that otherwise scattered members experience themselves as a group for, in reality, the group exists nowhere but in the minds of its isolated members. Accordingly, they have no place and no time to experience (and thereby re-create) the group but during those ritual occasions when the members assemble, engage in the so-called sacred rituals (whose value of **sacredness** is, for Durkheim, simply the product of the group's collective behavior and thus focus, not an expression of some inner quality in an act or an object), and leave confident of their identity. Durkheim therefore concludes that religion is the name given to a collection of social behaviors and social institutions; God-talk is, in fact, group members symbolically talking about an ideal sense of the group itself. This analysis of the social function of *la vie religieuse* (the religious life, as he phrased it), then, is rather different from prior and subsequent **essentialist** scholars, either **theologically** essentialist or, as in the Intellectualists, **naturalistically** essentialist. In fact, the speculations on timeless origins (such as **E. B. Tylor**'s work on the origins of **animism**) would strike a Durkheimian scholar as untestable (since time travel does not exist) and therefore unscientific.

> 'If religion protects man against the desire for self-destruction, it is not that it preaches the respect for his own person to him with arguments *sui generis*; but because it is a society. What constitutes this society is the existence of a certain number of beliefs and practices common to all the faithful, traditional and thus obligatory. The more numerous and strong these collective states of mind are, the stronger the integration of the religious community, and also the greater its preservative vale. The details of dogmas and rites are secondary. The essential thing is that they be capable of supporting a sufficiently intense collective life. And because the Protestant church has less consistency than the others it has less moderating effect upon suicide.'
>
> — *Suicide: A Study in Sociology* (1897)

> 'Primitive classifications are therefore not singular or exceptional, having no analogy with those employed by more civilized peoples; on the contrary, they seem to be connected, with no break in continuity, to the first scientific classifications. In fact, however different they may be in certain respects from the latter, they nevertheless have all their essential characteristics... Their object is not to facilitate action, but to advance understanding, to make intelligible the relations which exist between things. Given certain concepts which are considered fundamental, the mind feels the need to connect to them the ideas which it forms about other things. Such classifications are thus intended, above all, to connect ideas, to unify knowledge; as such, they may be said without inexactitude to be scientific, and to constitute a first philosophy of nature.'
>
> — *Primitive Classification* (1903; co-written with Marcel Mauss)

'A society can neither create nor recreate itself without creating some kind of ideal by the same stroke. This creation is not a sort of optional extra step by which society, being already made, merely adds finishing touches; it is the act by which society makes itself, and remakes itself, periodically... A society is not constituted simply by the mass of individuals who comprise it, the ground they occupy, the things they use, or the movements they make, but above all by the idea it has of itself... Therefore, the collective ideal that religion expresses is far from being due to some vague capacity innate to the individual; rather, it is in the school of collective life that the individual is able to conceive of the ideal.'

— *The Elementary Forms of Religious Life* (1912)

Diana L. Eck

Diana Eck, a Professor of **Comparative Religion** and Indian Studies at Harvard University, graduated from Harvard University with her PhD in 1976. Although her early work was devoted to religion in India, she has increasingly devoted her attention to issues of religious pluralism, advocating tolerance, mutual understanding, and acceptance of difference by means of **inter-religious dialogue**, especially as these all manifest themselves in contemporary US politics. She has been active in the United Methodist Church, the World Council of Churches and, since 1991, has been the Director of The Pluralism Project. This collaborative project, funded initially through the Lilly Endowment and now also funded by the Ford and Rockefeller Foundations, coordinates a network of sixty local affiliates, involving approximately 100 scholars working on affiliated projects that chronicle the changing shape of religious diversity both in the US and elsewhere in the world.

Eck on Religion

As with others in both the liberal **theological** and **humanistic** traditions, Eck understands religion to concern a domain of personal **experience** of deep meaning and significance thought to be uniformly shared by people despite their different historical, geographic and **cultural** locations. Her preference for such terms as '**experience**' and '**faith**' indicates the priority her work gives to understanding religion as an inner sentiment that defies categorization and which is expressed publicly in such historically conditioned forms as narrative (**myth**), practice (**ritual**), and institutional systems (traditions) – all of which are often considered static by many of their participants but which are, Eck argues, the constantly changing and growing outer form of a uniform inner faith. Thus, acts of categorization (such as the attention given to distinguishing denominations from each other, let alone the distinction between different religions and between religion and other aspects of daily life) are themselves part of the difficulty to be overcome, along with overcoming differences among the merely secondary manifestations of this faith.

> 'Tolerance is a deceptive virtue. I do not wish to belittle tolerance, but simply to recognize that it is not a real response to the challenging facts of difference. Tolerance can enable coexistence, but it is certainly no way to be good neighbors. In fact, tolerance often stands in the way of engagement. If as a **Christian** I tolerate my **Muslim** neighbor, I am not therefore required to understand her, to seek out what she has to say, to hear about her hopes and dreams, to hear what

it meant to her when the words "In the name of Allah, the Merciful, the Compassionate" were whispered into the ear of her newborn child.'

— *Encountering God: A Spiritual Journal from Bozeman to Banaras* (1993)

'America's religious diversity is here to stay, and the most interesting and important phase of our nation's history lies ahead. The very principles on which America was founded will be tested for their strength and vision in the new religious America. And the opportunity to create a positive multireligious society out of the fabric of a democracy, without the chauvinism and religious triumphalism that have marred human history, is now ours.'

— *A New Religious America: How a 'Christian Country' Has Become the World's Most Religiously Diverse Nation* (2001)

Mircea Eliade (1907–1986)

Throughout much of the mid- to late twentieth century there was no more influential scholar of religion than Mircea Eliade, the Romanian expatriate. After attaining some fame in Romania as a novelist after World War I, Eliade spent the World War II years abroad, and wrote books in the late 1940s and early 1950s for which he would later become famous throughout the world – volumes on **comparative religion**, shamanism, and yoga. In the late 1950s he held a brief visiting appointment at the University of Chicago's Divinity School and, following the unexpected death of the program's then Chair – the German sociologist of religion, Joachim Wach – Eliade stayed on and, along with the scholar of Japanese religions, Joseph Kitagawa, played a central role in leading Chicago's program to a place it continues to hold as one of the field's most important graduate programs. Eliade was classically trained as a comparativist and is today best known for his efforts to establish what at Chicago is called 'History of Religions' as an autonomous, academic discipline, distinct from **anthropological, psychological** or **sociological** studies of religion. His largely successful approach to accomplishing this, adopted by others both before and after him, was to argue for the *sui generis* nature of religion, thereby requiring distinct methods for its study and distinct institutional locations for carrying out this research. Because of the unique character of religious phenomena (each being the site where 'the **sacred**' manifests itself), along with his views that religion was at its **essence** concerned with establishing meaning in otherwise potentially meaningless human lives and societies, Eliade was also known for his advocacy of what he termed a total **hermeneutics** (that is, the study of religion being a complete interpretive science of human beings, what might be called 'the Queen of the Sciences', a term once reserved for theology), what he also called the New **Humanism**; the historian of religions, by studying symbolic expressions of what he held to be deeply meaningful existential situations common to all peoples, was able to re-experience in their own lives – and thereby become the interpreters of and guardians for – the meaning that these symbols, narratives and practices once had for archaic peoples long ago. Apart from a tremendously impressive amount of writing and editing (including his role, toward the end of his life, as the editor-in-chief of what has become the field's primary reference work, *The Encyclopedia of Religion* [1987]), Eliade is also known today for the manner in which, after his death in 1986, his life (some of its details were made public through his four published volumes of journals and his two-volume autobiography) and his extensive body of work have generated a substantial body of critical secondary literature, concerned with re-examining his arguments in favor of religion's irreducible character as well the way in which – like many European intellectuals who matured between the two World Wars – his personal politics may have impacted his scholarship.

Eliade on Religion

Eliade argued that an essential component of all human beings was their need to make their worlds meaningful, which was carried out through their interconnected systems of symbols, **myths** and **rituals** – all of which provided human beings with orientation in an otherwise chaotic world of historical existence (which implies a linear movement from a known past to an utterly unknown, and therefore terrifying, future). One could say, then, that the human condition, according to Eliade, was coming to grips with what he termed the 'terrors of history'. Hence Eliade's interest in studying tales of cycles and returns (the myth of the eternal return), belief systems involving rebirth, geography and architecture oriented toward a center (as in a central tent pole), and rituals that marked a point as the center of the village or the world (Latin: *axis mundi*, the central pivot point of the world or the entire universe). What he termed *Homo religiosus* (Latin, religious man) was best exemplified, he believed, in archaic or primitive peoples, since for them – unlike **modern**, secular people – the cosmos was entirely sacred; nonetheless, even secular people have no choice but to create meaning, so they too shared this (sometimes suppressed) aspect with their archaic counterparts, making them also attuned to the times and locations where meaning ruptured into the otherwise ambiguous historical world, thereby providing humans with a point of reference, a center point. Such points – what he termed hierophanies (from Greek *hiero*, meaning holy; a showing or a manifestation of the Sacred) – could be anything, from a rock to a tree, from the moon, to the tides or a mountain, even movies and literature, not to mention the so-called traditional elements of religion such as pilgrimage, worship, fasting, etc. The collection of symbols, narratives, practices and institutions we call religion were, for Eliade, the preeminent site where meaning-making took place, ensuring that he saw any approach to the study of religion that was not hermeneutical (such as the explanatory analysis of **reductionists**) to be highly problematic for it 'explained away' the very thing he thought to be of most importance; because such meanings are camouflaged, only the careful interpreter could uncover them.

> '[A] religious phenomenon will only be recognized as such if it is grasped at its own level, that is to say, if it is studied as something religious. To try to grasp the essence of such a phenomenon by means of physiology, psychology, sociology, economics, linguistics, art or any other study is false; it misses the one unique

and irreducible element in it – the element of the sacred... Because religion is human it must for that very reason be something social, something linguistic, something economic – you cannot think of man apart from language and society. But it would be hopeless to try and explain religion in terms of any one of those basic functions which are really no more than another way of saying that man is. It would be as futile as thinking you could explain *Madame Bovary* by a list of social, economic, and political facts; however true, they do not affect it as a work of literature.'

— *Patterns in Comparative Religion* (1949)

'The sacred always manifests itself as a reality of a wholly different order from "natural" realities... The first possible definition of the sacred is that it is the opposite of the **profane**. Man becomes aware of the sacred because it manifests itself, shows itself, as something wholly different from the profane. To designate the act of manifestation of the sacred, we have proposed the term hierophany... It could be said that the history of religions – from the most primitive to the most highly developed – is constituted by a great number of hierophanies, by manifestations of sacred realities.'

— *The Sacred and the Profane: The Nature of Religion* (1959)

'For the historian of religions the fact that a myth or a ritual is always historically conditioned does not explain away the very existence of such a myth or ritual. In other words, the historicity of a religious **experience** does not tell us what a religious experience ultimately is. We know that we can grasp the sacred only though manifestations which are always historically conditioned. But the study of these historically conditioned expressions does not give us the answer to the question: What is the sacred? What does a religious experience actually mean?'

— 'The "Origins" of Religion', in *The Quest: History and Meaning in Religion* (1969)

James G. Frazer (1854–1941)

Born in Glasgow, Scotland, and educated at the University of Glasgow with a second baccalaureate at Trinity College at Cambridge, James Frazer is a noteworthy British **anthropologist** and historian and is among a group of late nineteenth-century scholars known as Intellectualists (a group generally interested in using **evolutionary** theory to study early humans in light of their mental abilities). He was also concerned with studying cultural phenomena through the lens of the comparative method, assuming that collecting descriptive information on many variations of, for instance, a particular institution, would help to shed light on the origins of the institution. Looking solely within one **culture** was therefore not sufficient to explain the origins and role of actions or beliefs. Instead of doing fieldwork himself, Frazer (like all early anthropologists) relied on the letters, journals, and manuscripts from missionaries, explorers and military personnel (indicating the intimate, though sometimes unrecognized, link between European **colonial** expansion, on the one hand, and gains in scientific knowledge, on the other). Frazer wrote, in his highly influential book *The Golden Bough*, about the behaviors of the participants in a Hellenistic ritual and then compared their actions to that of modern 'primitives'. *The Golden Bough* examined an ancient **ritual** that is said to take place in the city of Aricia, near Rome. The ritual involves a priest whose duty it was to guard the grove near Lake Nemi. If a slave happened to escape from his master and kill the priest, he would win his freedom but also take on the responsibility of guarding the grove. Frazer was fascinated by this story for its insights into what he called the 'primitive mind'. Through his research Frazer hoped to shed light on the current behaviors of those involved in 'primitive' religions by using his method of comparative studies. Frazer's thesis (similar to that fellow Intellectualist **E. B. Tylor**'s) was that the mental capacities of humankind developed in the same evolutionary way as did the human body. Further, Frazer hypothesized that magic was the behavioral predecessor of religion just as religion was the intellectual predecessor of science, therefore, modern 'primitives' could have much in common with the classical Hellenistic mind. Frazer has played a major role in the study of religion through his impact on modern scholars. Although *The Golden Bough* grew into a vast, multi-volume work that few people might read in its entirety today, it is one of the first scholarly books that consistently employed the comparative method.

Frazer on Religion

Throughout *The Golden Bough*, Frazer builds a **theory** of magic (the effort to manipulate, in a non-physical manner, objects in the historical world) and its role in the eventual formation of **religion**, and subsequently what we today know as science. For Frazer, magic (the attempt to manipulate events in the natural world) can be divided into the two categories of homeopathic and contagious magic, both of which are contained under the heading of sympathetic magic. Sympathetic magic is based upon what Frazer refers to as the Law of Similarity and the Law of Contact. Homeopathic magic is understood by the magician to function under the Law of Similarity. The magician uses his magic to produce certain situations by mimicking their behaviors. An example would be a magician who pours water on the ground to induce rain, or someone who manipulates a doll expecting the same actions to be performed on the actual person. Contagious magic, on the other hand, is based on the Law of Contact. Frazer believed that magicians who used this kind of magic believed that items had great associations with one another, and even after the items had been separated from their source they could still have an effect on each other. The example that Frazer gives for this type of magic is when a portion of a person's body is severed; there remains the belief that the body part still has a connection to the body such that if something is done to the part the person will feel the results regardless of the physical distance from the estranged hand, foot, hair, etc. By classifying and studying magic in this way, Frazer aimed for his work to be applicable to many different cultures, all of which were thought to use these various types of magic. Moreover, he argued that magic was the evolutionary precursor to religion (which involved the belief that supernatural beings could be persuaded to influence events in the world), and that all cultures that now have religion must have had a previous period of magic. Furthermore, Frazer argued that societies that now possess science must have gone through previous periods of religion and magic. By applying the biological evolutionary theory to the study of societies Frazer was also able to hypothesize about some of the 'primitive' groups with whom nineteenth-century European travelers had come into contact. In fact, Frazer believed that the Australian Aborigines could be understood by means of his theory. Along with many of his peers, Frazer believed that they were the most primitive culture in existence at that time; they were thought to have been virtually frozen in time, and therefore to have no form of religion whatsoever (given earlier, largely **Christian**-influenced definitions of religion as 'belief in a God and an afterlife'; of course this view has largely been discredited), thereby relying exclusively on magic in their attempts to influence the world around them.

'Along with the view of the world as pervaded by spiritual forces, savage man has a different, and probably still older, conception in which we may detect a germ of the **modern** notion of natural law or the view of nature as a series of events occurring in an invariable order without the intervention of personal agency. The germ of which I speak is involved in that sympathetic magic, as it may be called, which plays a large part in most systems of superstition... In short, magic is a spurious system of natural law as well as a fallacious guide of conduct; it is a false science as well as an abortive art. Regarded as a system of natural law, that is, as a statement of the rules which determine the sequence of events throughout the world, it may be called Theoretical Magic; regarded as a set of precepts which human beings observe in order to compass their ends, it may be called Practical Magic.'

'Wherever sympathetic magic occurs in its pure unadulterated form, it assumes that in nature one event follows another necessarily and invariably without the intervention of any spiritual or personal **agency**. Thus its fundamental conception is identical with that of modern science; underlying the whole system is a faith, implicit but real and firm, in the order and uniformity of nature.'
— *The Golden Bough: A Study in Magic and Religion* (1890)

Sigmund Freud (1856–1939)

Although no longer considered at the forefront of theoretical developments in the field of **psychology**, Sigmund Freud nonetheless remains
important as the father of psychoanalysis. Along with the work of **Karl
Marx** and **Emile Durkheim**, his research on the interactions between
individual and group has contributed to a field today known as social
theory. Born in 1856 in Freiberg, Moravia (today a region in the Czech
Republic), his family later moved to Vienna, the city where he would
spend the rest of his life. Trained as a medical doctor with an interest in
neurology, he was forced to abandon the medical profession when his
method of treating patients by means of hypnosis was deemed unscientific by his colleagues. Wanting to develop a more scientific approach to
the study of the mind, he applied principles from the natural sciences,
especially physics, yet he concluded that the complexities of the mind
required more sophisticated and comprehensive explanations. To develop
such **theories**, he studied, among other things, his patients' reports of
their dreams. Freud theorized that human minds not only have a conscious component but also an unconscious aspect, the content of which
manifests itself when the conscious mind is not in control, such as in
dreams, fantasies, and most importantly, in neuroses (that is, abnormal
behaviors such as those classified as obsessive compulsive disorders or
uncontrollable fears of such things as water or public places). His psycho-
analytic theory names the individual components of the human psyche
as: the id (Latin for 'it' meaning the unconscious and uncontrollable primal instincts), the superego (Latin for 'I above' meaning those social influences from the outside world that are imposed upon the human per-
sonality from birth [making toilet training a fascinating moment for some
Freudians to study since it is among the earliest moments when the social
group forcibly imposes its will on the young individual]), and the ego
(Latin for 'I' meaning the mediator between the superego and the id). For
Freud, the inevitable competition between what he termed the pleasure
principle (embodied by the id's drive for self-gratification) and the reality
principle (embodied by the superego's self-policing activities) was the
primary cause of neuroses in the human psyche. Society, with its rules
and laws, was one of the main sources of censure; repression of the
pleasure principle/id – which he deemed to be instinctual, primal, and
the source of uncontrollable though natural urges and desire – was therefore the basis of social life. Because all humans are both biological individuals with natural needs and desires as well as actors in society, Freud
concluded that each human needed to engage in repression and thus
possessed some form of neurosis.

Freud on Religion

Although Freud's theory initially dealt with the individual, he eventually included society in his studies, such that **myth** functioned on the social level as dreaming did for the individual. Freud argued that a **culture**'s myths, fairy tales, art, legends, **rituals**, etc., were manifestations of society's collective psyche; religion being a site where socially dangerous urges and desires could be expressed in socially harmless ways. Freud therefore identified religion as one of the main sites of conflict and re-pression for human beings. He explained that religion was an illusion, something we wish to be true, which helped humans to cope with feel-ings of helplessness, weakness, and the inability to gratify the self in all instances – much like the fantasies we know as dreams help the indi-vidual to cope with antisocial desires which can therefore not be acted upon in reality. Freud's theory of religion especially applied to the Judeo-Christian **worldview** that consists of a patriarchal god-figure. For Freud this 'father-figure' god represents a childlike faith in the biological father's ability to protect us. Therefore, religion functions as a protective device to help humans cope with a hostile physical reality which daily frustrates their natural desires. Yet, at the same time, we experience a love/hate relationship with our 'father-figure' because their nurturing and protective capacity also places rules and limitations upon the id. Much as humans are conflicted in their feelings toward authority figures, so too they are conflicted toward their gods – which are, he concluded, merely symbols of actual authority figures. It should therefore be clear that Freud's theory understands religion in a non-***sui generis*** manner; his work shifts the focus from identifying some essential core element to studying religion's function as a coping mechanism for individuals living within social groups.

> 'In view of these resemblances and analogies one might venture to regard the obsessional neurosis as a pathological counterpart to the formation of a religion, to describe this neurosis as a private religious system, and religion as a universal obsessional neurosis. The essential resemblance would lie in the fundamental renunciation of the satisfaction of inherent instincts, and the chief difference in the nature of these instincts, which in the neurosis are exclusively sexual, but in religion are of egoistic origin.'
>
> — 'Obsessive Acts and Religious Practice' (1907)

> 'That the effect of religious consolations may be likened to that of a narcotic is well illustrated by what is happening in America. There they are now trying – obviously under the influence of petticoat government – to deprive people of all stimulants, intoxicants, and other pleasure-producing substances, and instead, by way of compensation, are surfeiting them with piety. This is another experi-ence as to whose outcome we need not feel curious.'
>
> — *The Future of an Illusion* (1927)

'Religion is an attempt to master the sensory world in which we are situated by means of the wishful world which we have developed within us as a result of biological and psychological necessities.'
— *New Introductory Lectures in Psychoanalysis* (1933)

Clifford Geertz (1926–2006)

Clifford Geertz is among the best known and most influential US **anthropologists** of the mid- to late twentieth century. Geertz is known especially among scholars of religion for his often utilized definition of **religion** as, in his famous words, 'a **cultural** system'. Having obtained his PhD from Harvard University in 1956 – after serving in the US Navy during the last years of World War II – Geertz held academic positions at the University of California, Berkeley, University of Chicago, Oxford University, and Princeton University's prestigious Institute for Advanced Studies. His early studies of Javanese culture (Java is an island that is part of the Indonesian archipelago and which contains the country's capital, Jakarta) were followed by repeated fieldwork – now generally understood as a requirement for producing legitimate anthropological knowledge. Geertz spent time in such other places as Bali and Morocco, ensuring that his work has been particularly well known to some scholars of modern **Islam**. Unlike his anthropological predecessors who were intent on explaining the natural causes of elements of culture, Geertz is today associated with an anthropological tradition known as symbolic anthropology, which is concerned with studying the meaning (as opposed to either the origins or the causes) that beliefs, behaviors, institutions and symbols have for the members of a culture – which is itself seen by those who follow Geertz as an elaborate, interconnected system of symbols. As such, Geertz is part of the **hermeneutic** and **phenomenological** traditions of scholarship – traditions with a long and still active history in the **humanistic** study of religion. That is, to study a culture adequately, one must understand fully the meaning that a system of symbols and actions has for a group of cultural actors; this understanding presupposes correct interpretation on the part of the observer. In a classic example Geertz borrowed from the British philosopher Gilbert Ryle (1900–1976), he cites an observer – inevitably disconnected from the 'local knowledge' possessed by a group of cultural participants – witnessing what might be called a 'wink'. Yet, this observer is incapable of distinguishing a meaningless twitch from a sly wink from what could even be an elaborate parody of a wink (in which the secrecy sometimes communicated by a wink is undermined by being broadcast to the entire group). For those who possess this knowledge, this seemingly subtle body movement could mean anything from an attempt to lessen the bite of a criticism to a recognition that someone was 'in on the joke', to a sexual advance. To understand the meaning of the wink – and therefore to understand the manner in which shared sets of interconnected ideas and symbols (that is, cultures) make our worlds

inhabitable by making them meaningful and therefore sensible to us –
required what Geertz famously described as 'thick description' of culture's
interconnected symbols and the changing contexts in which they operate. A merely 'thin description' of the behavior known as 'a wink' was
therefore hardly sufficient to understand it.

Geertz on Religion

For scholars of religion, Geertz's best known work is surely his 1966 essay
'Religion as a Cultural System'. This long essay is in fact simply a definition of religion accompanied by a detailed commentary. As promised by
the title of the essay, Geertz defined religion as a system of meaningful
symbols that function to establish interior dispositions (what he termed
'moods and motivations') which assisted people to build what some might
call a **worldview** which was understood to be authentic and authoritative.
Despite his interest in studying the meaning of these symbols as experienced by the participants, Geertz nonetheless clearly advocated an anthropological approach to the study of religion as a human institution, a
sub-system within culture as a whole. This having been said, he nonetheless assumed, along with many contemporary scholars of religion, that
what we call religion constituted an inner world – not necessarily one of
faith, as advocated by various **theologians**, but one of affectations and
sentiments (that is, 'moods and motivations'). Geertz is also known for
his two notions of experience-near and experience-distant (coined by the
Viennese psychoanalyst, Heinz Kohut [1913–1981], they are roughly comparable to the technical terms **emic/etic**). Because non-participants do
not necessarily share these inner moods and motivations (signified by the
terms above) it is incorrect to assume a strict division between the participant and the non-participant; instead, somewhat akin to a **family resemblance** approach to definition, he argued that people are more or less
familiar/estranged from various **experiences** and situations. The goal of
anthropological fieldwork is therefore to find analogies and points of contact to assist the observer, who does not share the mood and for whom
some symbolic activity (such as a wink) may appear insignificant, with
transforming a new and therefore experience-distant moment, symbol or
situation into one that is near, familiar, and therefore understandable.
This exercise is premised on the assumed basis of the commonality of the
human condition and the universality of religion – a presumption for which
Geertz has been strongly critiqued by the anthropologist **Talal Asad**.

> 'Religion is sociologically interesting not because, as vulgar **positivism** would
> have it, it describes the social order (which, in so far as it does, it does not only
> very obliquely but very incompletely), but because, like environment, political

power, wealth, jural [legal] obligations, personal affection, a sense of beauty, it shapes it... For an anthropologist, the importance of religion lies in its capacity to serve, for an individual or for a group, as a source of general, yet distinctive, conceptions of the world, the self, and the relations between them, on the one hand – its model of aspect – and of rooted, no less distinctive "mental" dispositions – its model for aspect – on the other. From these cultural functions flow, in turn, its social and psychological ones.'

— 'Religion as a Cultural System' (1966; reprinted in *The Interpretation of Cultures: Selected Essays* [1973])

'An experience-near concept is, roughly, one that someone – a patient, a subject, in our case an informant – might himself naturally and effortlessly use to define what he or his fellows see, feel, think, imagine, and so on, and which he would readily understand when similarly applied by others. An experience-distant concept is one that specialists of one sort or another – an analyst, an experimenter, an ethnographer, even a priest or an ideologist – employ to forward their scientific, philosophical, or practical aims. "Love" is an experience-near concept, "object cathexis" [Greek *kathexis*, meaning a holding] is an experience-distant one. "Social stratification" and perhaps for most people in the world even "religion" (and certainly "religious system") are experience-distant; "caste" and "nirvana" are experience-near, at least for **Hindus** and **Buddhists**.'

— '"From the Native's Point of View": On the Nature of Anthropological Understanding' (1974; reprinted in *Local Knowledge: Further Essays in Interpretive Anthropology* [1983])

David Hume (1711-1776)

The Scottish philosopher, David Hume, was born in Edinburgh, studied at the University of Edinburgh, and, upon graduation in 1725, intended to practice law. However, his interest in philosophy, political theory, **history** and literature soon became his focus. While in France in the mid 1730s, he wrote his *A Treatise of Human Nature*, and throughout the late 1730s and early 1740s he wrote on such topics as moral theory and politics. Denied an academic position at the University of Edinburgh in 1745 (due to his growing reputation as a skeptic), Hume took a position as the secretary to a British Army general, traveling throughout France, and, over the course of the next decade, published books for which he is still famous today: *Philosophical Essays Concerning Human Understanding* and *An Enquiry Concerning the Principles of Morals*. Hume held positions as a librarian and had an appointment at the British embassy in Paris, though he resigned from his government position in 1769.

Hume on Religion

Although still studied by philosophers and political theorists, in the academic study of religion Hume is, perhaps, best known as the author of two works that contributed much to establishing the basis for a naturalistic theory of religion: *A Natural History of Religion* (1757) as well as his *Dialogues Concerning Natural Religion* (1779; written in the dialogical style, somewhat similar to a dialogue by Plato, in which speakers representing distinct philosophical viewpoints investigate a topic). Hume made a break with a **theological** approach to the study of religion, asking not what it meant to be religious or inquiring in the proper way of being religious but, instead, enquiring into the causes of religious beliefs. Although his work was carried out early in the history of this alternative approach – evidenced in such things as Hume's apparent assumption that a creator obviously exists – he stands out from his contemporaries for his interest in explaining the natural causes of human knowledge about God. Observing that religious belief is not universal (and that even when it is found it differs dramatically), Hume reasons that the cause of such beliefs cannot be something innate or **essential** to **human nature**; instead, he argues that beliefs in supernatural powers must itself be caused by something more basic. He finds this basis in what he terms the 'hopes and fears' of human beings who have no choice but to live in a world in which the future is unknown and the actual causes of events are often unknown. Speculating on the **experiences** of early human beings interacting with a sometimes threatening natural world he concludes that the belief in a powerful, superhuman being controlling events is a device human beings have long used (much as **anthropomorphism** is such a device) to make sense of their inhospitable environments. Religion, thus, is not basic to human nature; instead, Hume helps to lay the basis for a **naturalistic** approach by arguing that knowledge about the gods can be explained by **reducing** it to 'hopes and fears'.

'We are placed in this world, as in a great theatre, where the true springs and causes of every event are entirely concealed from us; nor have we either sufficient wisdom to foresee, or power to prevent those ills, with which we are continually threatened. We hang in perpetual suspense between life and death, health and sickness, plenty and want; which are distributed amongst the human species by secret and unknown causes, whose operation is oft unexpected, and always unaccountable. These unknown causes, then, become the constant object of our hope and fear.'

'There is a universal tendency among mankind to conceive all beings like themselves, and to transfer to every object, those qualities, with which they are familiarly acquainted, and of which they are intimately conscious. We find human faces in the moon, armies in the clouds; and by a natural propensity, if not corrected by experience and reflection, ascribe malice or goodwill to every things, that hurts or pleases us... No wonder, then, that mankind, being placed in such an absolute ignorance of causes, and being at the same time so anxious concerning their future fortune, should immediately acknowledge a dependence on invisible powers, possessed of sentiment and knowledge.'

— *The Natural History of Religion* (1757)

William James (1842–1910)

Older brother to the famous US novelist Henry James, William James attained fame of his own, in North America as well as Europe, as a **psychologist** and as an early **theorist** of religion. Educated as a young man in Europe, James received his medical degree from Harvard in 1869, taught anatomy and physiology there, established an experimental psychology lab, was the first to teach psychology in a US university (1875), and within a few years was also lecturing in philosophy. By the time he was invited, for 1901–2, to Edinburgh University, Scotland, to deliver its famous lecture series (established in 1888 by Lord Gifford, the Gifford Lectures continue to this day to 'promote and diffuse the study of Natural Theology in the widest sense of the term – in other words, the knowledge of God'), James had already become a noted philosopher of religion, publishing in 1897 a collection of ten essays entitled *The Will to Believe* (some dating to the late 1870s on such topics as morality and faith). Topics that had preoccupied him up until this point became the topic of his Gifford Lectures, published the following year under the title, *The Varieties of Religious Experience*. Drawing on his work in psychology, James focused on the various types of religious **experience** that, according to him, predated any expression of religion as found in narrative, behavior, and institution. Unlike the early **naturalistic theorists** of his time, James makes clear in these still very famous lectures that religious experience is not a mistaken apprehension of some element in the natural world, distorted by consciousness, but is, instead, a unique sort of experience not to be dismissed or explained away; the **theology** in these lectures is therefore most evident as is his defense of religious **faith** – found in his earlier writings – from the explanations of what was at that time called medical **materialism**. Today, James is also remembered as an early advocate of pragmatism – the philosophical view, prominent among some US philosophers, that, according to James's interpretation, beliefs are tested by the observation of their consequences.

James on Religion

James is perhaps best known today for his famous definition of religion, found in his 1901–2 Gifford Lectures, which makes clear that he was concerned to study religion as a subjective, individual phenomenon, rather than as part of a social system. In fact, for James social life was part of a secondary, external world in which one's primary feelings and experiences were expressed, always inadequately and to the detriment of the feeling itself (as evidenced in the common phrase, 'I can't quite put it into words'). Accordingly, the religious person in her or his solitude – and most importantly, the so-called 'religious genius' whose experiences initiate and animate followers and, eventually, entire social movements called religions – ought to be the proper object of focus when studying religion, for everything else is merely 'second-hand religion', as James the **idealist** would phrase it. James goes on to distinguish between two types of religious experience (and thus systems built upon these experiences): the 'healthy-minded' (those optimists whose outlooks exclude considering the existence of evil) and the 'sick-souled' (those whose outlooks take into account the reality of evil). Although the former works to a point, James concludes that only the latter is capable of adequately providing a basis for existing in the world, since evil must be confronted and accounted for. In addressing mysticism James makes evident his pragmatic bent, noting that mystical experiences are to be judged not medically (in terms of, for example, delusional states) but in terms of their results; 'mystical states', he concludes, 'may, interpreted in one way or another, be after all the truest of insights into the meaning of life.'

'Religion ... shall mean for us the feelings, acts, and experiences of individual men in their solitude, so far as they apprehend themselves to stand in relation to whatever they may consider the divine. Since the relation may be either moral, physical, or **ritual**, it is evident that out of religion in the sense in which we take it, theologies, philosophies, and ecclesiastical organizations may secondarily grow. In these lectures, however, as I have already said, the immediate personal experience will amply fill our time, and we shall hardly consider theology or ecclesiasticism at all.'

'The word **"religion"**, as ordinarily used, is equivocal. A survey of **history** shows us that, as a rule, religious geniuses attract disciples, and produce groups of sympathizers. When these groups get strong enough to "organize" themselves, they become ecclesiastical institutions with corporate ambitions of their own. The spirit of politics and the lust of dogmatic rule are then apt to enter and to contaminate the originally innocent thing; so that when we hear the word "religion" nowadays, we think inevitably of some "church" or other; and to some persons the word "church" suggests so much hypocrisy and tyranny and meanness and tenacity of superstition that in a wholesale undiscerning way they glory in saying that they are "down" on religion altogether. Even we who belong to churches do not exempt other churches than our own from the general condemnation. But in this course of lectures ecclesiastical institutions hardly concern us at all. The religious experience which we are studying is that which lives itself out within the private breast.'

'Knowledge about a thing is not the thing itself... If religion be a function by which either God's cause of man's cause is to be really advanced, then he who lives the life of it, however narrowly, is a better servant than he who merely knows about it, however much. Knowledge about life is one thing; effective

occupation of a place in life, with its dynamic currents passing through your being, is another. For this reason, the science of religions may not be an equivalent for living religion; and if we turn to the inner difficulties of such a science, we see that a point comes when she must drop the purely theoretical attitude, and either let her knots remain uncut, or have them cut by active faith.'
— Lectures 14, 15, and Conclusion in *The Varieties of Religious Experience: A Study in Human Nature* (1902)

Gerardus van der Leeuw (1890–1950)

Gerardus van der Leeuw is today one of the best examples of an early to mid-twentieth century scholar applying some of the methods of philosophical **phenomenology** to the study of religion, conceived as something distinct from **theology**. As with many of his – and even subsequent – academic generation of religious studies specialists, he began with the study of theology, earning a Doctor of Theology degree at the University of Leiden, in the Netherlands, in 1916, with a dissertation on the gods of ancient Egypt. After working briefly as an ordained minister in the Dutch Reformed Church, van der Leeuw was appointed in 1918 to a newly created position in the **History of Religions** at the University of Groningen – a position that also entailed teaching Liturgy [Greek *leitourgia*, meaning public service to the gods; the study of how to carry out the proper rituals of worship]. Arrested briefly by the Germans in 1943, during their occupation of Holland, he later served as the first post-World War II Dutch Minister of Education, and, shortly before his death in 1950 was elected the first president of the International Association for the History of Religions (IAHR) – which remains the primary international organization of scholars of religion. Although the phenomenological method is still largely employed in the field – despite a number of criticisms of (i) the presumption that someone's **experiences** can be understood by another and (ii) the presumption that it is sufficient to study something merely 'as it presents itself', without inquiring into its natural causes – today, van der Leeuw's work is likely read mostly as an example of an early attempt to distinguish the study of religion as a **cultural**, **historical** practice from long-established theological studies that sought to assess the adequacy of each religion and religious practice. Given his life-long interest in **Christian** theology and the phenomenology of religion, the success of establishing this distinction has been questioned by commentators.

van der Leeuw on Religion

In his search for the **essence** or irreducible inner structure of all religion – which he, like many of his contemporaries, thought was particularly well represented in what was often called 'the primitive mentality', information which was gained through ethnographies of others – van der Leeuw settled on the term 'power', rather than **Rudolf Otto**'s more common term 'the holy' or, after him, **Mircea Eliade**'s 'the **sacred**'. For van der Leeuw, power is the more fundamental and therefore cross-culturally useful term, capable of naming the subjective experience of both early and modern peoples, to which they each then assign various names and qualities, along with methods for recognizing, acquiring and exchanging it. The set of beliefs, practices and institutions we name as religion (each of which requires its own phenomenological classification, such as mysticism, sacrifice, worship, soul, church) are, therefore, for van der Leeuw those aspects of culture that operate together to assign the most basic and all-encompassing significance to things by placing them in relation to what the participant sees as a total system of the whole, or the universe – what he characterizes near the end of his 1933 book, *Religion in Essence and Manifestation* (the original German edition was titled, *Phänomenologie der Religion*), as 'the last word' that is neither uttered out loud nor ever fully understood. Although this closing section of his book – entitled the 'Epilegomena' [Greek *epi*, meaning at, on, upon, besides + *legein*, to speak, to declare: a saying besides or a supplemental discourse] – provides one of the most systematic and therefore useful statements of the phenomenological method, it also contains a number of almost poetic, theological conclusions. Despite his efforts to understand religion as a cross-cultural universal – efforts far different from many of his more traditional theological peers in the European academy – in the end van der Leeuw's phenomenological method is more intent on developing a theology of **world religions** rather than engaging in their historical study.

> '[W]hen we say that God is the Object of religious experience, we must realize that "God" is frequently an extremely indefinite concept which does not completely coincide with what we ourselves understand by it. Religious experience, in other terms, is concerned with a "Somewhat". But this assertion often means no more than that this "Somewhat" is merely a vague "something"; and in order that man may be able to make more significant statements about this "Somewhat", it must force itself upon him, must oppose itself to him as being Something Other. Thus the first affirmation we can make about the Object of Religion is that it is a highly exceptional and extremely impressive "Other".'

> 'Phenomenology is the systematic discussion of what appears. Religion, however, is an ultimate experience that evades our observation, a revelation which in its very essence is, and remains, concealed. But how shall I deal with what is thus ever elusive and hidden? How can I refer to "phenomenology of religion" at all?... [P]henomenology knows nothing of any historical "development" of religion, still less of an "origin" of religion. Its perpetual task is to free itself from every non-phenomenological standpoint and to retain its own liberty...'
> — *Religion in Essence and Manifestation* (1933)

Bruce Lincoln

After obtaining his PhD in 1976 at the University of Chicago, studying the **History of Religions** under the direction of, among others, **Mircea Eliade**, Bruce Lincoln held an appointment at the Center for Humanistic Studies at the University of Minnesota before returning to the University of Chicago where he is today the Caroline E. Haskell Professor of the History of Religions. As with many classically trained Historians of Religions trained during the height of Chicago's influence in the field, Lincoln's work emphasizes the acquisition of ancient languages and a focus on texts to study **myth** and **ritual**; his data is derived from broad **historical** and **cultural** areas: from ancient Iran and India to Native American traditions, Norse mythology, the colonial era in Africa, and the Spanish revolution of the late 1930s. However, unlike many of his peers, he is interested in studying cultural practices as elements of systems in which power and privilege are being contested (rather than studying symbols, myths and rituals as the phenomena that are merely public expressions of **essential**, deep meanings). As such, the influence of Marxist social theory is apparent in his work, as is the role played by **discourse** analysis (as associated with the field of **semiotics**).

Lincoln on Religion

Unlike classical Marxists **theorists** of religion, Lincoln studies religious discourse not so much as an opiate, dulling working-class sensibilities and thereby oppressing them, but as the name given to one among many rhetorical strategies upon which groups – whether dominant or marginal – routinely draw to normalize their various claims to authority. This approach has led him to focus his scholarship on how social boundaries are contested, inverted and legitimized, through narratives, behaviors, and the manipulation of symbols. This focus enables him to move well beyond the traditional data examined by previous generations of scholars of religion – apparent in his study of how authority is contested and reproduced at a variety of historical and social sites. Although much of his work continues to comprise a close reading of texts in their context (in the tradition of his earlier work on the ideology of the **Hindu** text known as the *Veda*), Lincoln is also widely known for studying the social and political contexts of scholarly texts on religion and myth, demonstrating the manner in which scholarship itself is the product of national and class interests. Most recently, Lincoln has entered the realm of public discourse by writing a series of op-ed pieces in US daily newspapers, in which he examines the techniques and the geo-political effects of contemporary political rhetoric.

'The same destabilizing and irreverent questions one might ask of any speech act ought be posed of religious discourse. The first of these is "Who speaks here?", i.e., what person, group, or institution is responsible for a text, whatever its putative or apparent author. Beyond that, "To what audience? In what immediate and broader context? Through what system of mediations? With what interests?" And further, "Of what would the speaker(s) persuade the audience? What are the consequences if this project of persuasion should happen to succeed? Who wins what, and how much? Who, conversely, loses?"... When one permits those whom one studies to define the terms in which they will be understood, suspends one's interest in the temporal and contingent, or fails to distinguish between "truths", "truth-claims", and "regimes of truth", one has ceased to function as historian or scholar. In that moment, a variety of roles are available: some perfectly respectable (amanuensis, collector, friend and advocate), and some less appealing (cheerleader, voyeur, retailer of import goods). None, however, should be confused with scholarship.'

— 'Theses on Method', in *Method & Theory in the Study of Religion* (1996)

'It would be nice to begin with a clear and concise definition of "myth", but unfortunately that can't be done. Indeed, it would be nice to begin with any definition, but to do so would be misleading, it would undercut and distort the very project I intend to pursue. For in the pages that follow I will not attempt to identify the thing myth "is"; rather, I hope to elucidate some of the ways this word, concept, and category have been used and to identify the most dramatic shifts that occurred in their status and usage.'

— *Theorizing Myth: Narrative, Ideology, and Scholarship* (2000)

Burton L. Mack

Now retired from the Claremont School of Theology, in California, Burton Mack carried out his doctoral studies in Germany and has played a leading role in helping the modern field of New Testament studies reinvent itself as the historically-grounded field of Christian origins. The texts of the earliest **Christians** are therefore of relevance to Mack neither for the meaning they convey nor for their accuracy in depicting the origins of the movement, but because they are understood as artifacts (or better put, subsequent copies of long lost originals) from a series of particular **historical** worlds out of which a social movement began and grew. Although many contemporary scholars of religion studying the New Testament continue to do so in a traditional manner (engaging in **hermeneutical** studies), Mack helped to pave the way for current studies which examine the texts as evidence of self-perceived marginal groups contesting social boundaries and experimenting with alternative ways of building social identity in the turn-of-the-era Greco-Roman world. As such, the texts are understood by Mack as **myths** – not in the sense of lies or innocently fanciful tales but in the sense of narratives that reflect and advance specific ways of representing the world and, along with it, one's place in it. For example, his study of the Gospel of Mark concludes that one would be mistaken to read it as a historical narrative that can be judged accurate or not; instead, the text comprises a myth of origins conducive to the interests and needs of its writer and his community. Mack's work therefore also closely examines non-canonical texts (texts from the same era as those subsequently included in the Bible but which early Christian leaders excluded from their authoritative collections, or canon [from the Latin for rule or measuring line]) as well as the '**Q**' document. Such non-canonical resources are useful for those attempting to gain information on the earliest forms of the social movement rather than simply reading authoritative texts which portray the origins of the movement as later generations understood it to have taken place.

Mack on Religion

For those who consider writing the history of a religion to require special methods capable of communicating their **essence** and enduring meaning and value, Mack's work is likely controversial for it locates the beginnings of early Christianity within a mundane, but no less interesting, Hellenistic social world. Yet for those who understand Christianity to be a social movement like any other – with a variety of beginning points and a complex history of efforts to unify group members' perceived identities, interests, and their representations of both – Mack's work is welcomed as an attempt to account for the rise of this movement in a way that does not take for granted the historical accuracy of participants' own attempts to represent its origins (bringing to mind issues associated with the insider/outsider problem). Seeing the earliest texts as data in need of analysis, Mack brings a number of social theoretical tools to his readings, concluding that the texts provide evidence of their authors' attempts to construct and legitimize particular social worlds in which their group's interests could be accomplished. When it comes to questions concerning the identity of the historical Jesus (as opposed to what **theologians** might term the 'Jesus of faith'), Mack concludes that we might be able to understand this historical actor better if we use as our model the itinerant (that is, wandering) cynic teacher easily found in this era, in this part of the world (Cynicism names a loosely organized yet influential turn-of-the-era Hellenistic philosophical movement characterized by its critique of wealth, emphasis on living a modest and virtuous life, as well as specific types of persuasive argumentation). This model, Mack argues, is helpful in understanding Jesus and the early Jesus movements, especially in light of his reading of Q as having originally contained a collection of wisdom sayings attributed to Jesus. It should therefore be clear that, for Mack, religion is the name given to a collection of social and rhetorical techniques and institutions that accomplish what he refers to as mythmaking and social formation – the intertwined means by which the conditions favorable to collective life are made possible and reproduced over time and place. As such, he draws on the work of a variety of theorists in his studies of the New Testament, from **Karl Marx** and **Emile Durkheim** to **Jonathan Z. Smith** and a number of current **anthropological** theorists. Mack's own influence among scholars both within and outside the field of Christian origins (an example of the former would be **William Arnal**), therefore, can be linked to his willingness to suspend the common assumption that the texts we come to know as scriptures comprise a distinct domain or a special case, requiring unique methods for their study; instead, his scholarship has worked toward studying such texts as a social theorist would study any text.

> 'Once upon a time, before there were gospels of the kind familiar to readers of the New Testament, the first followers of Jesus wrote another kind of book. Instead of telling a dramatic story about Jesus' life, their book contained only his teachings. They lived with these teachings ringing in their ears and thought of Jesus as the founder of their movement. But their focus was not on the person of Jesus or his life and destiny. They were engrossed with the social program that was called for by his teachings. Thus their book was not a gospel of the Christian kind, namely a narrative of the life of Jesus as the Christ. Rather it was a gospel of

Jesus' sayings, a "sayings gospel". His first followers arranged these sayings in a way that offered instruction for living creatively in the midst of a most confusing time, and their book served them well as a handbook and guide for most of the first Christian century. The book was lost ... to history somewhere in the course of the late first century when stories of Jesus' life began to be written and became the more popular form of charter document for early Christian circles.'

— *The Lost Gospel: The Book of Q and Christian Origins* (1983)

'All peoples tell stories about their past that set the stage for their own time and place in a larger world. This world expands the horizons of memory and imagination beyond the borders of their contemporary world and becomes populated with images, agents, and events that account for the environment, set precedents for social relations and practices, and intrude upon the daily round in odd and surprising ways. These agents and images usually have some features that are recognizably human, but are frequently combinations of features that do not normally appear in the real world and they can also be grotesque. Most people have not found it necessary or even interesting to reflect on the "truth" of their stories or grade them according to their degree of fantasy as has been the case in modern Western societies. When asked about such things by modern ethnographers, the answers have been a smile and a frown. As a story-teller for the Hopi Indians of the southwestern United States said, when asked how he knew their stories were true, "Because they are told".'

— *The Christian Myth: Origins, Logic, and Legacy* (2002)

Martin Marty

Although now retired from the University of Chicago's Divinity School, where he taught in the area of American religious history since 1963, Martin Marty continues to be active in the field. Quite apart from his role in training new generations of American religionists and his many publications, Marty has a national US presence through his many media appearances as an interpreter of issues that fall broadly within the area of contemporary US religion and politics. He is an ordained minister in the Evangelical Lutheran Church and is the past president of the American Academy of Religion, the American Society of Church History, and the American Catholic Historical Association. He has served on US Presidential commissions and has received many honorary doctoral degrees – all indications of the influential role his work has played in the late twentieth-century US academy. Along with R. Scott Appleby, Marty co-directed The Fundamentalism Project – a multi-year, collaborative research project under the auspices of the American Academy of Arts and Sciences and funded by the John D. and Catherine T. MacArthur Foundation. In 1998 the University of Chicago's Institute for the Advanced Study of Religion was renamed The Martin Marty Center (currently under the directorship of **Wendy Doniger**) – to which Marty regularly writes a web column for its feature, 'Sightings'.

Marty on Religion

Although he has written a number of descriptive, historical works on such **Christian** theological topics as baptism and the Lord's Supper, as with many recent scholars of US religious history, the perennial problem of 'the one' and 'the many' – how national uniformity is possible despite **cultural** plurality – has been a recurring theme in much of Marty's recent writings. For many such writers, the history of the United States of America holds a special place in their efforts to investigate how to build a pluralistic society, for the history of the US is easily understood as one of migration and thus a blending of differing populations, each bringing with them a host of competing values. Defining religion broadly as any belief system that addresses issues of meaning and purpose – that is, any issue taken by groups to involve what **Paul Tillich** termed an 'ultimate concern' – Marty argues that religions do not simply build group identity but make possible a sense of well-being and thus a sense of community. Taking religion into account – that is, taking religion seriously as an **essential** element of **human nature**, engaging not in **reductive** explanation but, instead, in conversation and **inter-religious dialogue** – is therefore key for anyone hoping to tackle the problems faced by pluralistic societies. It is precisely this concern that animates the multi-volumed publications of The Fundamentalism Project – an effort to document the manner in which some religious groups have responded to the twentieth century's spread of **modernity** (that is, the rapid success of specific forms of economic and political organization).

> 'America ... helped patent freedom for choice as well as freedom of choice in religion. As such, it can be seen as the first and most modern society. Today, as freedom and choice have both expanded and many citizens follow completely personal and private quests, we might even speak of an ultramodern phase in American pilgrimage.'
>
> — *Pilgrims in their Own Land: 500 Years of Religion in America* (1984)

> 'I do believe ... that the way to sort out the trivial from the urgent and the appropriate from the irrelevant is getting a variety of people together and starting a conversation. That's a technique suggested by a civil rights leader in Chicago more than three decades ago: "We just get a roomful of people", he explained, "and tell them not to come out until they have a solution". "To what problem?" "You'll find out quickly enough if you only start talking". So start talking.'
>
> — *Politics, Religion, and the Common Good: Advancing a Distinctly American Conversation About Religion's Role in Our Shared Life* (2000)

Karl Marx (1818–1883)

Although he was not primarily concerned with studying religion, as a political theorist Marx was interested in the social **function** religion played and how it made certain political and economic systems possible. Born in Prussia and originally trained as a philosopher, the young Marx turned from philosophy toward the study of economics and politics. In the early 1840s, he formed a life-long friendship with Friedrich Engels (1820–1895), with whom he co-wrote a number of his most famous works and who often financially supported Marx and his family. Historical **materialism** – the name given to Marx's **theory** of **history** – is based on the idea that the systems that organize and make possible human productive power (what he termed the modes of production) create the conditions in which human consciousness takes shape. As a materialist, Marx phrased it as follows: 'it is not the consciousness of men that determines their existence, but, on the contrary, their social existence that determines their consciousness'. His interest in **political economy**, therefore, had much to do with studying both systems of social rank and privilege, on the one hand, and systems of value and exchange, on the other, along with the types of thought and forms of identity they made possible. Marx paid particular attention to what he considered to be the harmful effects of the economic system known as capitalism; his critique was premised on his assumption that human labor ought to provide an opportunity to meet our inherent need for creative, fulfilling work. Inasmuch as capital (the profit that results from exchanging a product [what we could term a commodity] for more than it cost to produce) remains in the hands of those who own the production process (that is, those who own the means of production), and not necessarily in the hands of the person whose labor actually made the product (the worker), Marx concluded that workers in capitalist systems were exploited; he termed the working class 'the proletariat', from the Latin *proles*, meaning offspring; in ancient Rome, the *proletarius* was the lowest class of citizens. The proletariat, or wage laborers, do not own their own means of production (that is, they own no property and therefore have no access to accumulated wealth, or capital) and therefore have nothing to exchange but their own labor (that is, their bodies and their creativity) in order to make a living. The proletariat therefore live in a constant state of alienation – alienated from the results of their labor (whose purchase price had little to do with the wages they were paid to produce it) as well as from their own essentially creative **human nature** which has itself become little more than a commodity.

Marx on Religion

In his efforts to explain how oppressive political and economic conditions were perpetuated, and how people lived within such systems, Marx turned his attention to **religion** – understood by him as an **ideological** institution premised on the belief in a god and in an afterlife. According to Marx, religion – like all belief systems – is a product of material realities (i.e., who owns what) and thus a product of economic and political conditions. Thus, the problems of religion (such issues as salvation, suffering, redemption, punishment, guilt, etc.) are ultimately expressions of practical problems that exist within society; more explicitly, the problems of religion are merely a projection of problems with how social relations are organized. Religion – the belief in a better life to come – is therefore a symptom of oppressive social conditions. It is used by oppressors to distract people from the economic conditions in which they are forced to live and it is used by those who are exploited to cope with their lot in life – a form of coping with exploitation and alienation that, ironically, prevents them from ever changing the actual social conditions under which they live, for change is always removed from today to tomorrow, from this life to the next. Hence, for Marx, religion is famously identified as the 'opium of the masses' – it distracts and soothes people whose lives have been **reduced** to commodities. But by distracting and soothing them, by allowing them to put up with their lot in life, religion perpetuates the actual source of the problem – which lies in the realm of politics and economics, not **theology**.

> 'Man makes religion, religion does not make man. Religion is indeed man's self-consciousness and self-awareness so long as he has not found himself or has already lost himself again. But, man is no abstract being squatting outside the world. Man is the world of man – state, society. This state and this society produce religion, which is an inverted consciousness of the world, because they are an inverted world. Religion is the general theory of this world, its encyclopedic compendium, its logic in popular form, its spiritual *point d'honneur*, its enthusiasm, its moral sanction, its solemn complement, and its universal basis of consolation and justification. It is the fantastic realization of the human essence since the human essence has not acquired any true reality. The struggle against religion is, therefore, indirectly the struggle against that world whose spiritual aroma is religion.'

> 'Religious suffering is, at one and the same time, the expression of real suffering and a protest against real suffering. Religion is the sigh of the oppressed creature, the heart of a heartless world, and the soul of soulless conditions. It is the opium of the people.'

> 'The abolition of religion as the illusory happiness of the people is the demand for their real happiness. To call on them to give up their illusions about their

condition is to call on them to give up a condition that requires illusions. The criticism of religion is, therefore, in embryo, the criticism of that vale of tears of which religion is the halo.'

— *Contribution to the Critique of Hegel's Philosophy of Right* (1844)

Tomoko Masuzawa

After earning her PhD from the University of California at Santa Barbara in 1985, Masuzawa began her career at the University of North Carolina, Chapel Hill, where she was Associate Professor in the Department of Religious Studies and member of the Program in Social Theory and Cross-Cultural Studies. She currently holds a joint appointment in the Department of History and the Program of Comparative Literature, at the University of Michigan, where her course topics include European intellectual **history** and critical **theory**. Her research, which could be characterized as meta-theoretical, concentrates on the historical development of the nineteenth- and early twentieth-century search for the origin of religion and the history and politics of the categories **'religion'** and **'world religions'**.

Masuzawa on Religion

Unlike other scholars, Masuzawa is not concerned with studying religion as something that exists *sui generis* and she is uninterested in accumulating a wealth of descriptive, **phenomenological** information about religious beliefs, practices and institutions. Instead, in her most recent book, *The Invention of World Religions* (2005), she traces the practical implications of the European category of 'world religions' – as well as the category 'religion' itself – from its first appearance in the late eighteenth century to the contemporary usage of the category by anyone who employs it to classify and categorize certain sets of phenomena, as if these particular elements of culture are essentially unique and distinct from other aspects of social life. Masuzawa argues that the category – which is today employed as if it names an obvious aspect of **culture** – was first employed in an effort to move beyond conceptualizing religion in what has often been termed a Judeo-Christian sense (in which the world was once divided between the chosen or saved and heathens); instead, 'world religions' signals an attempt at a more pluralistic understanding (whose development coincided with the rise of European **colonial** influence, a crucially important point when tracing the history of this concept) that many other phenomena can be considered inherently religious. Accordingly, over time we see other groups included under this conceptual umbrella: **Islam** (once known throughout Europe as 'Mohammadism'), **Buddhism**, **Hinduism**, Sikhism, Taoism, **Confucianism**, Shinto, Native American religions, etc. Because not all cultures have produced social movements that easily fit the criteria of this concept, throughout history we have also seen the development of such sub-types as Primitive Religions, Archaic Religions, Traditional Religions, and (because these previous categories are now considered unusable) the now popular term, Indigenous Religions – all used to name groups that scholars, relying on the **family resemblance** approach, recognize as having some religion-like characteristics but which do not meet the criteria established for naming something as a world religion (e.g., geographic spread, influence, etc.). To illustrate the manner in which the assumptions that animate these criteria are rendered self-evident, Masuzawa draws attention to the standard map often located in a world religions textbook that illustrates the demographics of the major world religions. For her, this map legitimizes what is in reality a specific classification system of rather recent invention. Her work is important because, in showing the development of the category 'world religions', she exposes the mutable nature of all classification systems, rendering them accessible to historical analysis and critical study.

> 'What may be expected from this rereading – with, admittedly, a potentially insurgent intent – is, at least, gaining some ground, from which we may begin to challenge the position of the self-appointed "ascetic priest" (alias "**modern** scholar"), whose sanctimonious self-understanding seems to dictate unilaterally the entire configuration of the field of knowledge. What is sought here is a critique that departs fundamentally from the kind of vaguely narcissistic self-criticism within the confines of "ethics of science" – "Are we fair in our representations of the 'primitives'?" – which ultimately refuses to question the positional structure of the knowing and the known, and thus remains insistently blind to the question of power. What is called for instead is a critical inquiry about the practice of knowledge and power, about the politics of writing, as it pertains to the study of religion.'
> — *In Search of Dreamtime: The Quest for the Origin of Religion* (1993)

> 'In the **social sciences** and **humanities** alike, "religion" as a category has been left largely unhistoricized, **essentialized**, and tacitly presumed immune or inherently resistant to critical analysis. The reasons for this failing on the part of the academy, this general lack of analytic interest, and the obstinate opacity of the subject of religion, are no doubt many and complex. But the complexity may begin to yield to critical pressure if we are to subject this **discursive** formation as a whole to a different kind of scrutiny, a sustained and somewhat sinuous historical analysis.'
> — *The Invention of World Religion, or How European Universalism was Preserved in the Language of Pluralism* (2005)

F. Max Müller (1823–1900)

Friedrich Max Müller – a German-born scholar of the religions of ancient India and an early historian of **myth** and language (what was then called the field of Philology [the love of words, from the Greek *philos* + *logoi*]) who, after 1846, spent his professional life in England (first teaching at Oxford University and then working as the curator of its main research library) – is generally considered, along with his contemporary, the Dutch Egyptologist, Cornelius P. Tiele (1830–1902), as one of the nineteenth-century founders of what was then known as either **Comparative Religion** or the Science of Religion. Müller earned a PhD in philosophy, then moved on to study in Berlin and eventually Paris, where he continued his study of **Sanskrit** and comparative philology (the precursor to what we today know as **linguistics**). Like many of his generation, Müller was interested in the quest for origins, which, once determined, might account for what he saw as resemblances between members of what was then termed the 'Aryan' language family (also known as the Indo-Europeans – although this theory is today contested by some, it refers to a group hypothesized by some scholars to lie in the pre-history of many people in modern Europe and India; an ancient group whose presence was once thought to be detectable in the linguistic similarities among the ancient languages and in archeological evidence found throughout Greece, India, and Europe). Müller became famous for his multi-volume translations of ancient Sanskrit texts (some epics, others used in ritual ceremonies) that were previously unavailable to an English-speaking audience (notably, the collection of ancient chants or hymns known as the *Rig Veda* and, later in his career, overseeing the massive project published as *The Sacred Books of the East* [eventually consisting of fifty volumes]). He was also a well-known opponent of Darwin's theory of **evolution**, although he had his own views on the gradual development of myth and, eventually, religion, throughout human history. His interest in language and mythology – which led to work funded early on by the East India Trading Company, one arm of British **colonialism** – prompted him to argue for the comparative study of language and myth (which he understood as an inevitable 'disease of language', indicative of an early stage in human history), which, in turn, led to his interest in creating a new academic discipline: the scientific, cross-cultural study of religion as a basic aspect of **human nature** (much like language). In the mid- to late nineteenth century Müller became widely known in Britain for his highly successful public lectures and writings on the scientific and comparative study of religion, notably his programmatic book, *Introduction to the Science of Religion* (1873), and his collected essays, *Chips from a German Workshop* (the first of what were eventually four volumes was published in 1867).

Müller on Religion

Although today almost any bookstore will regularly stock a number of collections of world mythologies, it is crucial to note that, as recent as the mid-nineteenth century such collections were yet to come into existence. Instead, the work of collecting, translating and comparing texts only then being found and brought back to Europe by travelers, traders, missionaries, colonial bureaucrats, and members of the military fell to scholars such as Müller or **James Frazer**. Although such scholars were undoubtedly part of a colonial world, making it relatively easy to detect in their work unexamined assumptions regarding the superiority of European social organization and Christian forms of behavior and belief, it is to such scholars that we owe a debt for their painstaking archival work and their willingness to entertain that local and familiar practices are but one species of larger, cross-cultural patterns, shared by all human beings (thus making it possible to study not only any number of individual religions but also the collective human behavior common to them all, known as religion). So, quite apart from acting as a collector, translator and editor, Müller was also an early **theorist** of religion. His work emphasized the role played by myths (as with many, but not all, of his contemporaries, text and belief were prioritized over studying forms of behavior and organization) as the means whereby people were able to sense and communicate their sense of, what he called, 'the infinite' – a universal human intuition about the divine. Therefore, despite his role in helping to establish a scientific study of religion, his work understandably also shared something in common with a theological position which was in his day known as the study of natural religion – that is, taking the fact of religion for granted, it was the study of how humans gain knowledge of religion if they do not subscribe to a belief system that assumes this knowledge is revealed to them from another source, such as a god. At that time, 'natural religion' was therefore distinguished from 'revealed religion' – a basic classification, also apparent in **David Hume**'s earlier work, that was once used to distinguish among what were considered two different types of religion.

> 'When students of Comparative Philology boldly adopted Goethe's paradox, "*He who knows one language knows none*", people were startled at first, but they soon began to feel the truth which was hidden beneath the paradox. Could Goethe have meant that Homer did not know Greek, or that Shakespeare did not know English, because neither of them knew more than his own mother tongue? No! What was meant was that neither Homer nor Shakespeare knew what that

language really was which he handled with so much power and cunning. Unfortunately the old verb "to can", from which "canny" and "cunning", is lost in English, otherwise we should be able in two words to express our meaning, and to keep apart the two kinds of knowledge of which we are speaking. As we say in German, können is not kennen, we might say in English, to can, that is to be cunning, is not to ken, that is to know; and it would then become clear at once, that the most eloquent speaker and the most gifted poet, with all their command of words and skillful mastery of expression, would have but little to say if asked what language really is! The same applies to religion. He who knows one, knows none. There are thousands of people whose faith is such that it could move mountains, and who yet, if they were asked what religion really is, would remain silence, or would speak of outward tokens rather than of the inward nature, of the faculty of faith.'

— First Lecture in Lectures on the Science of Religion (1893)

'There is nothing more ancient in the world than language. The history of man begins, not with rude flints, rock temples or pyramids, but with language. The second stage is represented by myths as the first attempts at translating the phenomena of nature into thought. The third stage is that of religion or the recognition of moral powers, and in the end of One Moral Power behind and above all nature. The fourth and last is philosophy, or a critique of the powers of reason in their legitimate working on the data of experience.'

— Contributions to the Science of Mythology (1897)

Rudolf Otto (1869–1937)

Born in Piene, Germany, Rudolf Otto was one of the foremost German systematic **theologians** of the late nineteenth century. He was educated at the University of Erlangen and the University of Göttingen in liberal theology and history of the Bible. Although originally having planned on entering the ministry, Otto's arrangements were forced to change due to staunch resistance from the conservative German Lutheran Church and their hesitance to give him an appointment. Instead, Otto took a teaching position at the University of Göttingen and began studying the work of Jakob Friedrich Fries (1773–1843) – an influential German philosopher who worked to rationalize Immanual Kant's philosophy. Otto was so taken with Fries that he helped to begin a Neo-Fresian movement within his academic circle and wrote one of his first books on the philosophy of Fries and Kant. Otto is probably best known to scholars of religion for what is considered by most to be his best work, *The Idea of the Holy*. In this book, Otto contends that religion – or better put, religious experiences and sentiments – is a phenomena complete unto itself, or ***sui generis***. For this reason religion cannot be **reduce**d and thereby broken up into its constituent units, according to Otto (this supposition has come under scrutiny by modern scholars who disagree with Otto and instead use the theory of reductionism to provide insight into the nature of religion). Otto also thought that religion was knowable *a priori* (or independent of, or prior to, **experience**) and therefore its study comprises a completely different sphere of knowledge from other academic disciplines. Assuming this religious sentiment to be universal among human beings, Otto was also interested in the **history of religions** and traveled to India in 1911–1912 to study **Sanskrit** and **Hinduism**. It was through this journey that he began to struggle with the theological problems of the presumed **Christian** superiority in the face of his growing knowledge of what we now refer to as the '**world religions**'.

Otto on Religion

Although not necessarily read today to the extent that his work once was, Rudolf Otto's contribution to the study of religion has been tremendous and enduring; it can be attributed to his strongly argued thesis about the internal, participant-only, spiritual, non-empirical nature of religion. Otto argued that religion was *sui generis* and therefore a category completely unto itself. Religion – much as argued by **Friedrich Schleiermacher** before him, and **Mircea Eliade** after him – could therefore not adequately or wholly be understood through other disciplines like **psychology**, philosophy or **sociology**. Instead, it was only the individual who has had a distinctly 'religious' experience who could express what it is that characterizes religion. It is for this hypothesis that Otto can be classified with scholars who devote their time in the study of religion to the theory of **essentialism**. For Otto, there is one or many attributes that are contained within the experience we call 'religion' and all religions contain these elements although in varying degrees. Otto coined a term to name a category of feeling that he believed corresponded to a purely religious sentiment – the numinous (from the Latin *numen*, meaning a force or power often identified with natural objects; sometimes understood as meaning holy). The numinous, for Otto, was therefore a religious category of value that could be discussed, but could not be strictly defined because it was irreducible in nature. The feeling that an individual experiences, known as the experience of the numinous, is of something he termed the *mysterium tremendum et fascinans*. Otto defines this feeling as one that contains elements of utter awe, might, power, energy and urgency. The *mysterium tremendum et fascinans* – the significant, compelling (and thereby attractive), and yet repelling mystery of it all – is what religious participants experience while they are engrossed in religious ceremonies or in a particularly 'religious' state of mind. That each religious participant names the object of this experience differently, from religion to religion and from place to place, does not lessen what for Otto is the essence of these seemingly varied experiences.

> 'The reader is invited to direct his mind to a moment of deeply-felt religious experience, as little as possible qualified by other forms of consciousness. Whoever cannot do this, whoever knows no such moments in his experience, is requested to read no farther; for it is not easy to discuss questions of religious psychology with one who can recollect the emotions of his adolescence, the discomforts of indigestion, or, say, social feelings, but cannot recall any intrinsically religious feeling. We do not blame such an one, when he tries for himself to advance as far as he can with the help of such principles of explanation as he knows, interpreting "aesthetics" in terms of sensuous pleasure, and "religion" as a function of the gregarious instinct and social standards, or as something more primitive still. But, the artist, who for his part has an intimate personal knowledge of the distinctive element in the aesthetic experience, will decline his theories with thanks, and the religious man will reject them even more uncompromisingly.'

> 'The numinous ... issues from the deepest foundation of cognitive apprehension that the soul possesses, and, though it of course comes into being in and amid the sensory data and empirical material of the natural world and cannot anticipate or dispense with those, yet it does not arise out of them, but only by their means.'
> — *The Idea of the Holy: An Inquiry into the Non-Rational Factor in the Idea of the Divine and its Relation to Rationality* (1917)

Friedrich Schleiermacher (1768–1834)

Born in Breslau (the Germanized name for what is today the city of Wroclaw in Poland; as of 1871, it was part of the German Empire), Friedrich Daniel Ernst Schleiermacher was the son of a Prussian army chaplain, and is today remembered as a influential Protestant **theologian** who devised a manner to defend belief in God from the criticisms leveled by the skeptics of his day. He was educated initially in schools administered by the Moravian Church – a Reformation denomination that originated in the mid-fifteenth century in ancient Bohemia and Moravia (what is today part of the Czech Republic) that emphasized the role of piety (from the Latin, *pietas*), an inner **experience** of the Gospel's saving power, over dogma and the so-called trappings of **ritual** and institution. Against his father's wishes, Schleiermacher left a Moravian seminary in 1787 and, instead, moved to the University of Halle, in east central Germany. Founded in 1694, the University of Halle is considered to have been among the first so-called modern universities in which religious orthodoxy and Church control over the curriculum gave way to free rational inquiry. There, Schleiermacher was thoroughly schooled in, among other topics, philosophy, especially the work of influential Prussian philosopher, Immanuel Kant (1724–1804). Kant is today among a very small group of deeply influential writers from this period, remembered best for his attempts to bridge **David Hume**'s arguments in favor of empiricism (the position that sensory experience is the basis of knowledge) with those of rationalism (the position that the innate ideas, or 'categories', of human reason, not experience, are the basis of knowledge). In 1794 Schleiermacher was ordained, then served as a hospital chaplain in Berlin, and went on to represent the Romantic movement (a philosophically **idealist** movement in the late eighteenth and early nineteenth centuries) by writing a number of important works that sought to defend religious **faith** against the attacks of Enlightenment skepticism (prompted both by empiricism and rationalism).

Schleiermacher on Religion

Although a Protestant theologian who made a significant contribution to the study of systematic theology – and thus certainly warrants the attention of those interested in the history of Protestant theology in Europe – Schleiermacher is remembered by scholars of religion for the manner in which he argued that faith (aside: he significantly employed the German word *Glaube*, meaning faith or justified belief, in the title of his major work, *Der christliche Glaube* [*The Christian Faith*]) operated not in the realm of reason but, instead, was akin to an aesthetic sense or feeling that could neither be supported nor criticized by reason. True religion, he therefore argued in his widely read *On Religion: Speeches to its Cultured Despisers* (first translated into English in 1893 and still in print) was a 'sense and taste for the Infinite'. His efforts to conceptualize religion as a private sentiment or intuition – efforts that were well in step with the manner in which previous generations of Protestant Reformers had successfully critiqued (and thereby subverted and eventually replaced) the authority of the Roman Catholic institution – continue to have influence today, most notably among writers who presume that religious practices, narratives and institutions are mere expressions of a presumably universal experience or faith. Because many such writers are intent on doing cross-cultural work, they do not necessarily follow Schleiermacher's lead in concluding that the object of this feeling is one's awareness of a complete and utter reliance (what he termed an 'absolute dependence') upon the **Christian** conception of God as revealed in the life of Jesus Christ, as communicated in the Gospels; nonetheless, many agree that the object of religious **discourse** is an affective (that is, emotional or aesthetic) sense that cannot be adequately grasped or studied by means of observation and rational argumentation. It is therefore quite possible, it would be argued, to be fully rational and religious at one and the same time.

> 'The usual conception of God as one single being outside the world and behind the world is not the beginning and the end of religion. It is only one manner of expressing God, seldom entirely pure and always inadequate... [T]he true nature of religion is neither this idea nor any other, but immediate consciousness of the Deity as He is found in ourselves and in the world.'
>
> — *On Religion: Speeches to its Cultured Despisers* (1799)

> 'But if anyone should maintain that there might be Christian religious experience in which the Being of God was not involved in such a manner, i.e., experiences which contained absolutely no consciousness of God, our proposition would certainly exclude him from the domain of that Christian belief which we are going to describe... [W]e assert that in every religious affection ... the God-consciousness must be present and cannot be neutralized by anything else, so that there can be no relation to Christ which does not contain also a relation to God... Just as there is always present in Christian piety a relation to Christ in conjunction with the God-consciousness, so in **Judaism** there is always a relation to the Lawgiver, and in Mohammedanism [i.e., **Islam**] to the revelation given through the Prophet.'
>
> — *The Christian Faith* (1820–1)

Ninian Smart (1927–2001)

Born in Cambridge, UK, Ninian Smart was classically trained at Oxford University in languages, **history** and philosophy, after first having served as a young man in Ceylon (now named Sri Lanka) in the mid- to late 1940s as a member of the British Army Intelligence Corps. But it was as a scholar of religion that he made his lasting international mark, notably at (among the many other universities at which he taught) the University of Lancaster, in the UK, and the University of California at Santa Barbara, in the US. Beginning in 1967 at Lancaster, and 1976 at UCSB, he played a pivotal role at both institutions in helping to establish thriving programs in the academic study of religion – a role that had much to do with not only his many writings on the proper method for conducting the public study of religion, as well as his well-known cross-cultural research on many of the **world's religions**, but also the long list of graduate students he trained throughout the years. To signify his tremendous impact on the international field, the Ninian Smart Annual Memorial Lecture was established after his death, with the location rotating each year between Lancaster and Santa Barbara. The first such lecture, delivered in Lancaster, was presented by **Mary Douglas** in 2002, followed by **Jonathan Z. Smith** in 2003. In 2005 the lecture was presented by **Wendy Doniger**.

Smart on Religion

Smart is, perhaps, best known today as a **phenomenologist** of religion. Many of his works – some of which are known to the contemporary student as world religions textbooks – were descriptive in nature, chronicling the traits that he argued comprised those things we commonly name as 'religions'. What he called his 'dimension **theory** of religion' named a collection of aspects or family of traits that typified religions. These dimensions – which include such traits as a narrative and behavioral component (that is, **myths** and **rituals**), an institutional component, and an aesthetic component – could, he argued, be found in many other human institutions; therefore, Smart favored using the broader term '**worldviews**' – thereby admitting nationalism, for example, to the group of phenomena studied – so as not to arbitrarily limit the scholar's work only to what we had traditionally known as religions. His early advocacy for what was once called a 'secular' study of religion, in contradistinction from a **theological** approach, placed him at the forefront of those who developed the modern institution known as Religious Studies, though he also retained an interest in such topics as **inter-religious dialogue**, which animated his late-in-life support for what he termed a World Academy of

Religion in which scholars of religion would interact with learned theologians from the world's many religious traditions.

'Many people, it is true, consider the very idea of looking at religion scientifically to be absurd or even distasteful. Absurd, because a scientific approach is bound to miss or distort inner feelings and responses to the unseen. Distasteful, because science brings a cold approach to what should be warm and vibrant. These hesitations about the enterprise are fundamentally mistaken, though understandable. They are mistaken precisely because a science should correspond to its objects. That is, the **human sciences** need to take account of inner feelings precisely because human beings cannot be understood unless their sentiments and attitudes are understood... As yet, the way in which one may deal with religion scientifically and, at the same time, warmly, is imprecisely understood... To return, however, to my opening paragraph, I am far from claiming that the study of religion is the most important thing to be undertaken in connection with religion. Being a saint is more important. But I would contend that, in the intellectual firmament, the study of religions is important not only because religions have been a major feature in the landscape of human life but also because a grasp of the meaning and genesis of religions is crucial to a number of areas of inquiry.'

— *The Science of Religion and the Sociology of Knowledge:
Some Methodological Questions* (1973)

'In providing a kind of physiology of spirituality and of worldviews, I hope to advance religious studies' theoretical grasp of its subject matter, namely that aspect of human life, **experience**, and institutions in which we as human beings interact thoughtfully with the cosmos and express the exigencies of our own nature and existence. I do not here take any faith to be true or false. Judgment on such matters can come later. But I do take all views and practices seriously... This implies that, in describing the way people behave, we do not use, so far as we can avoid them, alien categories to evoke the nature of their acts and to understand those acts... In this sense phenomenology is the attitude of informed empathy. It tries to bring out what the religious acts mean to the actors.'

— *Dimensions of the Sacred: An Anatomy of the World's Beliefs* (1996)

Jonathan Z. Smith

There is perhaps no more influential scholar of religion currently working than the University of Chicago's Jonathan Z. Smith – a widely published essayist and respected senior scholar who is also known for his strong commitment to undergraduate teaching and the place of the liberal arts curriculum in the modern university. Born in 1938 and raised in Manhattan, New York, Smith earned his BA in the late 1950s from Haverford College in Pennsylvania, and went on to earn his PhD from what was then Yale's newly established Department of Religion (which, in 1962, was instituted separate from Yale's Divinity School). Early in his career he worked briefly at the newly established religious studies department at the University of California, Santa Barbara, but soon joined the faculty of the University of Chicago in 1968, where he has remained throughout his career. Although the focus of his doctoral dissertation was **James G. Frazer**'s classic work on **myth** and **ritual**, *The Golden Bough*, it is evident that even at this early stage Smith was primarily concerned with the problem of method, that is, how to go about doing comparative work (an issue that has occupied his attention throughout his career). Since that time, much of Smith's data has derived from the religions of antiquity, including ancient **Judaism** and the earliest forms of **Christianity**, though his academic interests have taken his work to any number of different historical periods, languages and cultures – evident in his wide-ranging essays that often open by juxtaposing two seemingly unrelated pieces of data from human **history**, all in an effort to make evident a common theoretical point.

Smith on Religion

His ability to work in diverse data domains has more than likely led to J. Z. Smith's considerable influence among a wide number of scholars, many of whom find themselves drawn to his attention to detail, his unwillingness simply to assume that cultural difference is secondary to some presumed deep similarity that awaits detection, as well as his efforts to put before his colleagues the fact that scholarship (like **culture** itself) is always an act of choice, rather than an exercise in passively recognizing and then interpreting timeless meanings that lurk within texts, actions and symbols. This attention to choice places Smith's work at the heart of the field's recent turn toward emphasizing theory and the study of scholarship's motives and implications – for attention to the issue of selection presupposes one examine the criteria that are in use, which in turn presupposes that one examines the interests and goals that drive the selection process. Perhaps there is no better example of how this is put to work in Smith's writings than his well-known 1974 essay (first published in 1980 and which appears as chapter 4 in his influential collection of essays, *Imagining Religion*) 'The Bare Facts of Ritual'. Concerned with developing a way to study reports of ancient bear-hunting practices, Smith argues that rituals are parts of systems used by human communities to, as he puts it there, exercise an economy of signification or, as he writes in the closing lines of chapter 10 to his recent book, *Relating Religion: Essays in the Study of Religion* (2005), ritual ought to be understood as a 'character-istic strategy for achieving focus'. This implies that those routinized prac-tices that we know as rituals can be understood as one of the means by which groups direct their members' attention on things and priorities – which is another way of saying that they are a means whereby groups distract their members' attention from yet other things and priorities. The effect of common social practices that routinize focus is the reproduction of a specific sense of the group, as heading in a specific direction with specific goals, all premised on a shared sense concerning who does and does not count as a member and what counts as significant, memorable, understandable, and thus an item of knowledge. It is therefore in this sense that Smith might be considered an example of a modern day Intel-lectualist – if by this we no longer mean what we once might have by the term (as in a nineteenth-century **anthropological** movement) and, in-stead, signify merely those scholars who are interested in how human communities make sense of – make intelligible, meaningful and persua-sive – the worlds that they inhabit.

'If we had understood the archaeological and textual record correctly, man has had his entire **history** in which to imagine deities and modes of interaction within them. But man, more precisely western man, has had only the last few centuries in which to imagine religion. That is to say, while there is a staggering amount of data, phenomena, of human **experiences** and expressions that might be characterized in one culture or another, by one criterion or another, as religion – *there is no data for religion*. **Religion** is solely the creation of the scholar's study. It is created for the scholar's analytic purposes by his imaginative acts of comparison and generalization. Religion has no existence apart from the academy. For this reason the student of religion, and most particularly the **histo-rian of religion**, must be relentlessly self-conscious. Indeed, this self-consciousness constitutes his primary expertise, his foremost object of study. For the self-conscious student of religion, no datum possesses intrinsic interest. It is of value only insofar as it can serve as *exempli gratia* [an e.g.] of some fundamental issue in the imagination of religion. The student of religion must be able to articulate clearly why "this" rather than "that" was chosen as an *exemplum*. His primary skill is concentrated in this choice. This effort at articulate choice is all the more diffi-cult, and hence all the more necessary, for the historian of religion who accepts

neither the boundaries of canon nor of community in constituting his intellectual domain, in providing his range of *exempla*.'
                                        — *Imagining Religion: From Babylon to Jamestown* (1982)

'From the point of view of the academy, I take it that it is by an act of human will, through language and history, through words and memory, that we are able to fabricate a meaningful world and give place to ourselves. Education comes to life at the moment of tension generated by the double sense of "fabricate", for it means both to build and to lie. For, although we have no other means than language for treating with the world, words are not after all the same as that which they name and describe. Although we have no other recourse but to memory, to precedent, if the world is not forever to be perceived as novel and, hence, remain forever unintelligible, the fit is never exact, nothing is quite the same. What is required at this point of tension is the trained capacity for judgment, for appreciating and criticizing, the relative adequacy and insufficiency of any proposal or language and memory. What we seek to train in college are individuals who know not only that the world is more complex than it first appears, but also that, therefore, interpretative decisions must be made, decisions of judgment that entail real consequences for which one must take responsibility, from which one cannot flee by the dodge of disclaiming expertise.'
                        — 'The Introductory Course: Less in Better', in *Teaching the Introductory Course in Religious Studies: A Sourcebook* (1991)

Wilfred Cantwell Smith (1916–2000)

Born in Toronto, Wilfred Cantwell Smith graduated in 1938 with his undergraduate degree from the University of Toronto, studying **oriental** languages. He carried out **theological** studies in England working with, among others, the famous Islamicist, H. A. R. Gibb (1895–1971) – one of the editors of the famous multi-volume, *Encyclopedia of Islam*. During most of the years of World War II (1940–45), Smith was in India working with the Canadian Overseas Missions Council, teaching on such topics as the history of India and of **Islam**. (He was also ordained in 1944.) After the war he returned to school, earning his PhD in 1948 at Princeton University. Widely known for his early work on Islam, especially his commitment to cross-cultural comparison and the role played by empathy in one's studies, Smith is also known for his work on methodology (that is, his studies on how one ought to go about studying religions), his interest in developing a global theology of religious pluralism (premised on **inter-religious dialogue**), as well as his administrative work in helping to establish centers for pursuing the academic study of religion in general, or Islam in particular (e.g., at McGill University, in Montreal, at Harvard University, and at Dalhousie University, in Halifax).

Smith on Religion

Although his interests were clearly driven by theological assumptions, Smith is remembered as being among the first to study the history of the category '**religion**' – a term that, he argued (in the tradition of **Schleiermacher**), was inadequate because it is used to name what are, he argued, two utterly different things that ought not to be conflated: the outer, 'cumulative tradition', on the one hand, and, on the other, the inner **experience** of what he termed '**faith** in transcendence'. For Smith, it was the latter, this faith, that prompted various outward expressions that eventually came to be institutionalized and, because they were easily observed, came to be mistaken by scholars for the core of religion. In this regard his work could be characterized as an example of an **essentialist** approach to defining religion, insomuch as Smith presumed that the observable, public elements that we associate with such things as **ritual** and symbol were but derivatives of a prior, inner experience that was distinct from all other sorts of experiences. Moreover, assuming this faith to be the common core to all religion, Smith understandably developed considerable interest in working toward what he termed a global theology in which the differences among those public elements that he considered secondary and derivative could be overcome so as to bring about a cooperative pluralism among the **world's religions**, or, better put, the world's religious *traditions* – a term that Smith preferred because it focused attention on the traditions that built up around, but were not to be confused with, the faith that inspired them. In fact, it is in large part due to his influence that we today so commonly refer to religions as religious traditions.

> 'For I would proffer this as my second proposition: that no statement about a religion is valid unless it can be acknowledged by that religion's believers. I know that this is revolutionary, and I know that it will not be readily conceded; but I believe it to be profoundly true and important. It would take a good deal more space than is here available to defend it at length; for I am conscious of many ways in which it can be misunderstood and of many objections that can be brought against it which can be answered only at some length. I will only recall that by "religion" here I mean as previously indicated the faith in men's hearts.'
> — 'Comparative Religion: Whither and Why', in *The History of Religions: Essays in Methodology* (1959)

> 'It is customary nowadays to hold that there is in human life and society something distinctive called "religion"; and that this phenomenon is found on earth at the present in a variety of minor forms, chiefly among outlying or eccentric peoples, and in a half-dozen or so major forms. Each of these major forms is also called "a religion", and each one has a name: **Christianity**, **Buddhism**, **Hinduism**, and so on.
>
> I suggest that we might investigate our custom here, scrutinizing our practice of giving religious names and indeed of calling them religions. So firmly fixed in our minds has this habit become that it will seem perhaps obstreperous or absurd to question it. Yet one may concede that there is value in pausing occasionally and examining ideas that we otherwise take for granted.'
> — *The Meaning and End of Religion* (1962)

Herbert Spencer (1820–1903)

Born in Derby, England, Herbert Spencer was raised in an atmosphere of religious dissent and staunch individualism. During his childhood and adolescence, Spencer was influenced largely by the Quakers and the Unitarians of the Derby Philosophical Society. His father and uncle also held strong anti-clerical and anti-establishment views. Spencer was formally trained as a civil engineer but soon began to be interested in those intellectual pursuits that we today might term the **social sciences**. It was Spencer who first published a theory of **evolution** and coined the term 'survival of the fittest' – not Charles Darwin as many people today assume. Spencer's early works, such as *Social Statics, or the Conditions Essential to Human Happiness*, were concerned with the notion of civil liberties and the progression of human rights viewed through the lens of early evolutionary theory. Spencer's work was therefore largely influenced by his ideas on the evolution of human beings' physical body as well as their mind. In his largest work, *A System of Synthetic Philosophy*, Spencer applies his evolutionary theory to account for many aspects of human culture and its development over time. For instance, **human nature**, according to Spencer, is not contained within a group of **essential** characteristics; instead, it is based upon an ever changing and evolving set of social circumstances. Several of his volumes are included within *A System of Synthetic Philosophy*, which discusses such topics as biology, **psychology**, **sociology**, and ethics – all of which, Spencer believed, can be explained by appealing to one unifying theory (that of evolution) – a prime example of nineteenth-century **reductionism**.

Spencer on Religion

Much of Spencer's scholarly work was based on nineteenth-century scientific tenets and the theory of evolution – a theory that Spencer applied to the study of religion. Spencer was a firm believer in the necessity of empirical data; if a **theory** could not be supported or explained through the use of what Spencer determined as empirical facts about the world, then the theory was unknowable and speculation about it was unfounded. In holding this view, Spencer advocated a viewpoint commonly known as agnosticism. He believed that, although it was impossible to prove that the fundamental foundations of religion were wrong, it was equally impossible to prove that they were correct. Spencer therefore believed that it was of no use to debate such topics as the existence of God because it was beyond our scope of knowledge. Because of Spencer's **agnosticism** he was uninterested in a **theological** reading of religion. Instead, he chose

to approach the study of religion through such **naturalistic** fields as sociology, psychology and biology. In one of the volumes of *A System of
Synthetic Philosophy* (a volume entitled *The Principles of Sociology*) Spencer builds a thesis on the evolution of religious beliefs in humankind. He
begins by postulating that primitive human beings first formed a belief in
the supernatural through a system of magic, used to manipulate their
natural environment. After apparently seeing their long lost dead in dreams
and visions, Spencer postulates that primitive humans would have believed that these ghosts were ethereal manifestations of family members
and hence lead to ancestor worship. This ancestor worship was a predecessor to later 'higher' forms of organized religion as demonstrated by the
'higher races'. Although Spencer was once considered to be one of the
foremost authorities on the evolution of religious beliefs, in recent years
his work came under much criticism due to his views on 'primitives' and
their **cultures**.

> 'And now, we have prepared ourselves, so far as may be, for understanding
> primitive ideas. We have seen that a true interpretation of these must be one
> which recognizes their naturalness under the conditions. The mind of the sav-
> age, like the mind of the civilized, proceeds by classing objects and relations
> with their likes in past **experience**. In the absence of adequate mental power,
> there results simple and vague classings of objects by conspicuous likenesses,
> and of actions by conspicuous likenesses; and hence come crude notions, too
> simple and too few in their kinds, to represent the facts. Further, these crude
> notions are inevitably inconsistent to an extreme degree.'

> 'It is said, however, that ancestor-worship is peculiar to the inferior races. I have
> seen implied, I have heard in conversation, and I have now before me in print,
> the statement that "no Indo-European or Semitic nation, so far as we know,
> seems to have made a religion of worship of the dead". And the suggested
> conclusion is that these superior races, who in their earliest recorded times had
> higher forms or worship, were not even in their still earlier times,
> ancestor-worshipers... But that adherents of the **Evolution**-doctrine should ad-
> mit a distinction so profound between the minds of different human races, is
> surprising. Those who believe in creation by manufacture, may consistently hold
> that Aryans and Semites were supernaturally endowed with higher conceptions
> than Turanians... But to assert that the human type has been evolved from the
> lower types, and then to deny that the superior human races have been evolved,
> mentally as well as physically, from the inferior, and must once have had those
> general conceptions which the inferior still have, is a marvelous inconsistency.'
> — *The Principles of Sociology* (1899)

Rodney Stark

Although recognized as one of the leading contemporary US **sociologists** of religion, Rodney Stark initially studied journalism at the University of Denver and began his career as a reporter for the *Denver Post* in 1956. After a brief stint in the US Army, Stark enrolled at the University of California, Berkeley, and completed his PhD in sociology in 1971. From 1971 to 2003, Stark was professor of sociology and **comparative religion** at the University of Washington. Recently, he accepted an appointment as University Professor of the Social Sciences at Baylor University. Stark's extensive writing in the field of **Christianity**, which he has used as a domain to test his work in **rational choice theory** of religion, culminated in his book, *The Rise of Christianity*, which was nominated for the Pulitzer Prize in 1996.

Stark on Religion

Collaborating with William Sims Bainbridge, Rodney Stark proposed a series of **deductions** in order to uncover the two key components that both felt propelled religious participants: motives and exchanges. Using the model of rational choice theory, Stark and Bainbridge published *A Theory of Religion* (1987), in which they base the crux of their theory on seven basic axioms:

(1) Human perception and action take place through time, from the past into the future.
(2) Humans seek what they perceive to be rewards and avoid what they perceive to be costs.
(3) Rewards vary in kind, value, and generality.
(4) Human action is directed by a complex but finite information-processing system that functions to identify problems and attempt solutions to them.
(5) Some desired rewards are limited in supply, including some that simply do not exist.
(6) Most rewards sought by humans are destroyed when they are used.
(7) Individual and social attributes, which determine power, are unequally distributed among persons and groups in any society.

Employing the form of a deductive **theory** of religion, Stark and Bainbridge attempted to create a theoretical framework for the social-scientific study of religion. In his book, *The Rise of Christianity* (1996), Stark works within this sociological framework to argue that historically significant events (such as Emperor Constantine's conversion to Christianity) were secondary to the benefits realized through people's rational exchange of paganism, with its limited benefits, for Christianity, which was understood to hold greater promise of future benefits. For example,

Stark demonstrates that the poor treatment of women under Roman Law and culture significantly contributed to many women becoming believers in Christianity. By positing the Christian God as an 'exchange partner' capable of offering immense benefit to those who believe – both in this life and the one believed to come after – Rodney Stark further explicated the trajectory of polytheism to monotheism from the ancient world to the modern era in his more recent work, *One True God: Historical Consequences of Monotheism.*

> 'However, even if we use the best social science theories as our guide for reconstructing history, we are betting that the theories are solid and that the application is appropriate. When those conditions are met, then there is no reason to suppose that we cannot reason from the general rule to deduce the specific in precisely the same way that we can reason from the principles of physics that coins dropped in a well will go to the bottom.'
> — *The Rise of Christianity: A Sociologist Reconsiders History* (1996)

> 'The appropriate scientific assumption, and the one I have made every effort to observe, is **agnostic**: scientifically speaking, we do not know and cannot know whether, for example, the Qur'an was spoken to Muhammad by an angel or merely by his own inner voices. And, scientifically speaking, it doesn't matter! Our only access is to the human side of religious phenomena, and we can examine this with the standard tools of **social science**, without assuming either the real or the illusory nature of religion.'
> — *One True God: Historical Consequences of Monotheism* (2001)

Paul Tillich (1886–1965)

The German-born Paul Tillich was an ordained minister who is known today for his work in the US as one of the most influential Protestant systematic **theologians** of the early to mid-twentieth century. He studied at the Universities of Berlin, Tübingen, Halle and Breslau (where he was awarded his PhD in 1910), and served as an army chaplain during Word War I. Subsequent to that, Tillich held university appointments in Berlin, Marburg, Dresden and Frankfurt, though his position was terminated by the Nazi government in early 1933. By that Fall, Tillich had been invited to travel to the US to hold an appointment at Union Theological Seminary, in New York. Eventually, he also held appointments at Harvard University as well as the University of Chicago's Divinity School. Tillich's fame is the result of his efforts to create a theological system that took into account a series of early and mid-twentieth-century intellectual currents, including the influence of European **Existentialism**, the growing awareness, and thus interest, in cultures outside the Euro-North American world, as well as an interest in reconsidering the long-assumed split between **religion** and contemporary **culture**. Like many who have put their stamp on the field, he delivered the Gifford Lectures (at Scotland's University of Aberdeen), which resulted in one of the works for which he is best known today: the three-volume *Systematic Theology* (an effort to present a complete and coherent theological system). Tillich's normative scholarship (his interest in articulating the 'truth' and the 'meaning' of the **Christian** witness) distinguishes him from the modern study of religion, as does his attempt to define religion, which employs the common strategy of lodging religion within the individual by equating it with vague, subjective value judgments. Nonetheless, given the historical development of the academic study of religion from largely (Protestant) Christian theological concerns, Tillich can be seen as a transitional figure whose interest in contemporary culture, whose willingness to work with **Historians of Religions**, and whose efforts to understand religion 'in a wider sense', as he phrased it, prompted a generation of **humanistic** scholars to expand their interests to include cross-cultural analysis of religious symbols.

Tillich on Religion

The modern popularity among theologically-influenced scholars of religion of discussing 'religion and culture' (as opposed to conceiving of religion as but an item of or within human culture) owes much to Tillich's influence, as does the still widespread use of the category 'ultimate concern' when attempting to define religion. Regarding the first, Tillich betrays the influence of Existentialism – one of the most influential philosophical movements of his time – by attempting to create a Christian theology that bridged the gap between what he characterized as the unchanging infinite and the ever-changing finite. He did this by means of his retooled notion of 'Christ', which, for him, embodied the unification of enduring **essence** and **historical** existence (somewhat akin to early Christian attempts to articulate the presumed divine and yet human nature of Christ), thereby trying to avoid the contradiction between the two that was so apparent to Existentialist philosophers of his time. For Tillich, then, the conjunction 'and' in 'religion and culture' (a phrase that today names both entire departments as well as courses within many Religious Studies curricula) could, in a way, be understood as functioning in the manner of what he called 'Christ' – that which unites otherwise distinct spheres. In this attempt to overcome the critique of Existentialism, a *sui generis* understanding of religion is used, inasmuch as religion is thought to interact with culture, in specific and limited ways; to rephrase, in Tillich's system culture (and its constituent parts, such as political systems, economic practices, social structures, etc.) certainly does not cause religion; rather, religion is that which makes culture meaningful. Regarding Tillich's second lasting influence on the field, his proposal of defining the essence of religion as 'a dimension of depth', a **'faith** in an ultimate concern', has proved appealing to a number of scholars. Despite the obviously Christian nature of Tillich's work – evidenced in his use of 'God' for that which he also called an ultimate concern – the popularity of this definition of religion is linked to its apparent ability to be applied to a variety of social actors and cultural settings – for, unlike other theological attempts at definition, it does not merely emphasize faith in a particular sort of religious **experience**, symbol or institution but, instead, emphasizes 'God' as the basis of all Being. It is therefore thought by some to make it possible to hold the position that all human beings are religious, whether in conventional terms or not (a position he shared in common with his Chicago colleague, **Mircea Eliade**), for inasmuch as human beings are meaning-makers, they are presumed to have a faith in an ultimate concern that grounds and motivates their behaviors and commitments. Therefore, much as his notion of Christ expanded the concept from some of its previous understandings, so too his notion of religion pressed well beyond more traditional understandings that employed it to signify membership in specific sorts of institutions (e.g., shrines, mosques, synagogues, etc.) or belief in supernatural beings.

'Faith is a concept – and a reality – which is difficult to grasp and to describe. Almost every word by which faith has been described...is open to new misinterpretations. This cannot be otherwise, since faith is not a phenomenon besides others, but the central phenomenon in man's personal life, manifest and hidden at the same time. Faith is an essential possibility in man, and therefore its existence is necessary and universal... If faith is understood for what it centrally is, ultimate concern, it cannot be undercut by modern science or any kind of philosophy... Faith stands upon itself and justifies itself against those who attack it, because they can attack it only in the name of another faith. It is the triumph of the dynamics of faith that any denial of faith is itself an expression of faith, of an ultimate concern.'

— Dynamics of Faith (1957)

'Religious symbols are not stones falling from heaven. They have their roots in the totality of human experience including local surroundings, in all their ramifications, both political and economic. And these symbols can then be understood partly as a revolt against them... But what does this mean for our relationship to the religion of which one is a theologian? Such a theology remains rooted in its experiential basis. Without this, no theology at all is possible. But it tries to formulate the basic experiences which are universally valid in universally valid statements. The universality of a religious statement does not lie in an all-embracing abstraction which would destroy religion as such, but it lies in the depths of every concrete religion. Above all it lies in the openness to spiritual freedom both from one's own foundation and for one's own foundation.'

— 'The Significance of the History of Religions for the Systematic Theologian', in *The Future of Religion* (1966)

Edward Burnett Tylor (1832–1917)

Edward Burnett (E. B.) Tylor, one of the founders of the modern academic discipline of **anthropology**, belongs to a generation of academics known as the Intellectualists which includes **Müller**, **Spencer** and **Frazer**, all of whom helped pave the way for the modern academic study of religion. Raised and educated among Quakers (known also as the Society of Friends) and possessing no formal higher academic education, Tylor left his father's business in his early twenties and began his scholarly career doing fieldwork in the mid-1850s in Mexico under the guidance of the amateur British ethnologist (a scholar of cultural origins and functions) Henry Christy (1810–1865). In 1875, Tylor received an honorary doctorate from Oxford University where he was keeper of the Oxford University Museum (1883) and later became Britain's first (indeed, the first in the English-speaking world) Professor of Anthropology (1896), until his retirement in 1909.

Tylor on Religion

Tylor – who famously defined **culture** as 'that complex whole which includes knowledge, belief, art, morals, law, custom, and any other capabilities and habits acquired by man as a member of society' – held an **evolutionary** view concerning the development of culture and religion (sometimes also known as Social Darwinism), arguing that **animism** (belief in spiritual beings) was the earliest stage of what we today know as religious behavior. Despite his interest in what was then commonly known as 'primitive religion' (an interest motivated by the common nineteenth-century quest for the origins of religion), unlike some of his European contemporaries, who understood others as uncivilized savages, Tylor argued for a 'psychic unity of mankind', assuming instead that, despite differences in the stages of their evolutionary development, all humans (past and present) shared common **cognitive** functions (such as a curiosity to explain unexpected events in their environment). The goal of anthropological study, for Tylor, was therefore to develop a cross-culturally useful framework in which the evolution of culture could be explained and the nature of its origins understood.

> 'Scientific progress is at times most furthered by working along a distinct intellectual line, without being tempted to diverge from the main object to what lies beyond, in however intimate connexion... My task has been here not to discuss Religion in all its bearings, but to portray in outline the great doctrine of Animism, as found in what I conceive to be its earliest stages among the lower races of mankind, and to show its transmission along the lines of religious thought.'

'To the minds of the lower races it seems that all nature is possessed, pervaded, crowded, with spiritual beings. In seeking a few types to give an idea of this conception of pervading Spirits in its savage and barbaric stage, it is not indeed possible to draw an absolute line of separation between spirits occupied in affecting for good and ill the life of Man, and spirits specially concerned in carrying on the operations of Nature. In fact these two classes of spiritual beings blend into one another as inextricably as do the original animistic doctrines they are based on.'

— *Religion in Primitive Culture* (1873)

Max Weber (1864–1920)

Whereas the French sociologist **Emile Durkheim** has been influential on **reductionist** social theorists, the German **sociologist** and economist, Max Weber, has been just as influential on those scholars of religion who are part of what we could term the *Verstehen* (German, to understand, as in empathetically re-experiencing the feelings of another person) tradition which studies religion as a system of meanings (represented in part by the work of the US anthropologist, **Clifford Geertz**). Weber's work is therefore part of a tradition intent on understanding the meaning-worlds of the people scholars study. However, he has also been profoundly influential on scholars who argue for the value-free, or objective, nature of science in distinction to the subjective nature of value-judgments. Having studied law, **history** and **theology** early on, Weber earned his PhD from the University of Berlin in 1889 with a dissertation entitled 'The Medieval Commercial Associations' – a study of trading companies in medieval Italy and Spain. In the early to mid-1890s, he was a law professor at the University of Berlin and practiced law in Berlin as well. Taking a position at Freiburg University in 1894, Weber taught **political economy** and, in 1897, taught political science at Heidelberg University. However, after an ongoing nervous illness in the late 1890s and early 1900s, Weber left scholarship for a time, to return, from 1904 until his death, as a private scholar and editor, but without a university appointment (though he held a visiting appointment at the University of Vienna in the summer of 1917 and held an appointment in 1919 to the University of Munich). During the last fifteen years of his life, Weber edited an encyclopedia (*Foundations of Social Economics*), founded the German Sociological Society (1909), increasingly participated in public debates and journalism during the World War I years, participated in efforts to reform the post-War German government (along with being a member of the German Peace Delegation to Versailles, at the conclusion of the war), all the while producing what are today considered some of his most important cross-cultural and theoretical works on economics, ethics and religion.

Weber on Religion

Although he was certainly interested in studying the causes of people's behavior and not just describing their self-perceptions, Weber sought these causes not in the material conditions of people's lives – as did his predecessor **Karl Marx** – but instead, in their beliefs and ideas. Furthermore, to study these beliefs adequately, thereby allowing him to explain the believer's behaviors and the social systems that resulted, Weber advocated a **hermeneutical** method, aimed at understanding the meaning of these beliefs. His work therefore provides a complex blend of a number of positions in the study of religion that are often seen today to be in conflict. Perhaps the best example of Weber's method is provided by his effort to use differences in religious belief to account for the origins of the economic system that goes by the name of capitalism. Premised on the accumulation, and therefore concentration of capital (surplus money or wealth that can be reinvested) by means of owning the ability to make products that are exchanged for profit (that is, private ownership of the means of production, which leads to laborers being paid a wage that is less than the value for which goods are exchanged on the open market), the rise of capitalism struck Weber as being the direct result of certain world-denying behaviors (what we could call asceticism [from the Greek *askesis*, meaning practice or exercise]) specifically associated with the Protestant **worldview**. This view, he argued, was comprised of an individualistic focus in which the religious participant was thought to have a personal relationship with God mediated through their subjective experience of the Word of God; furthermore, in this worldview one was unaware of the eventual fate of one's soul. The so-called Protestant work ethic therefore resulted from a group's continual, disciplined reinvestment of its social energies and creativity regardless of the apparent profits that past investments had created, for it was believed that works (that is, the results of good deeds) alone did not provide evidence of one's salvation; only the grace of God, which works mysteriously by means of a criteria unknown to the devotee, can guarantee one's salvation (which identifies a key difference between traditional Protestant and Roman Catholic belief systems). The result of the devotee's unsure position in the world was a continued production of merit – either in the form of a social value that resulted from good behavior or monetary value that resulted from frugality. Accordingly, for Weber, beliefs about God, souls, and the afterlife – what comprises a very traditional understanding of religion – was a key factor in accounting for social behavior and the success of institutions. Neither Protestantism nor capitalism were represented as morally superior; rather, the former's belief system simply provided the necessary conditions for the development of the latter. Perhaps it is therefore understandable why Weber's work was (and remains) so controversial for those who wish to hold religion to be ***sui generis***; his linkage of religious belief to practical economics can strike some as demeaning to religion, given that it is assumed to be removed from (that is, above) the concerns of material life.

'A glance at the occupational statistics of any country of mixed religious composition brings to light with remarkable frequency a situation which has several times provoked discussion in the Catholic Press and literature, and in Catholic congresses in Germany, namely, the fact that business leaders and owners of capital, as well as the higher grades of skilled labor, and even more the higher technically and commercially trained personnel of modern enterprises, are overwhelmingly Protestant... The same thing is shown in the figures of religious affiliation almost wherever capitalism, at the time of its great expansion, has had a free hand to alter the social distribution of the population in accordance with its needs, and to determine its occupational structure... [T]he principles of explanation of this difference must be sought in the permanent intrinsic character of their religious beliefs, and not only in their temporary external historico-political situations.'

— *The Protestant Ethic and the Spirit of Capitalism* (1904–5)

'To define **"religion"**, to say what it is, is not possible at the start of a presentation such as this. Definition can be attempted, if at all, only at the conclusion of the study. The **essence** of religion is not even our concern, as we make it our task to study the conditions and effects of a particular type of social behavior. The external courses of religious behaviors are so diverse that an understanding of this behavior can only be achieved from the viewpoint of the subjective **experiences**, ideas, and purposes of the individuals concerned – in short, from the viewpoint of the religious behavior's "meaning".'

— *The Sociology of Religion* (1922)

Ludwig Wittgenstein (1889–1951)

Born in Vienna, Austria, Ludwig Wittgenstein was the youngest of eight children born into a wealthy family. He was trained as an engineer in Berlin, but later became interested in philosophy through the works of the acclaimed British philosopher, Bertrand Russell. Wittgenstein enrolled at Trinity College, Cambridge in 1912 and studied there until he enlisted in the Austrian army at the start of World War I. Although Wittgenstein's philosophical writings were heavily concerned with ethics, his place within the academic study of religion comes from other aspects of his work. His influence on the study of religion has mostly been due to his work on classification. He is responsible for questioning the ways in which human beings order their worlds and, in so doing, casting doubt on **monothetic** systems of classification. Wittgenstein argued that any **essentialist** notion of classification was fundamentally flawed. For him, the idea that collections of objects are related through sharing a common trait (essence) was nonsensical. Instead, he advanced the thesis that there are infinite spectrums of relations among like objects and that these varied relations are what cause objects to be grouped together. He concluded that it is precisely this **'family resemblance'** between objects that properly unifies systems of classification.

Wittgenstein on Religion

Applied to the study of religion – a field in which essentialist notions of religion's enduring significance, timeless origin, and inner meaning are widespread – Wittgenstein's work on classification and definition has had tremendous influence, resulting in a variety of family resemblance approaches to defining the delimited series of traits shared in common, to whatever degree, by those social movements we call religions. Like the cluster of partially overlapping circles known in mathematics as Venn diagrams, many scholars now see the family of religions as sharing any number of partial, non-essential traits. **Ninian Smart**, for one, identified a series of what he termed 'dimensions' which, as he phrased it, 'help to characterize religions as they exist in the world'. Among these traits are: the **ritual** dimension, the **experiential** dimension, the narrative dimension, the doctrinal dimension, the ethical dimension, the social dimension, and the material dimension (as in architecture). Of course, such a family resemblance approach to definition raises the question of how one judges the borderline cases, such as those who, along with Smart, wish to argue that Communism is a religion (or, better put, a **worldview**) or those who might wish to argue that sports such as baseball or American football qualify as a religion because they fit a sufficient number of the criteria by which other movements are judged to be religions (heightened sense of awe, building group identity, zones of the sacred distinguished from that of the profane, hallowed ancestors, etc.). If definitions are only as useful as they are limited (that is, 'pen' has utility because not everything gets to count as one), then how does one decide where to draw the line?

> 'Nothing we do can be defended absolutely and finally. But only by reference to something else that is not questioned. I.e. no reason can be given why you should act (or should have acted) like this, except that by doing so you bring about such and such a situation, which again has to be an aim you accept.'
> — *Culture and Value* (1980)

> 'Consider for example the proceedings that we call "games". I mean board-games, card-games, ball-games, Olympic games, and so on. What is common to them all? Don't say: "There must be something common, or they would not be called games" – but look and see whether there is anything common to all. For if you look at them you will not see something that is common to all, but similarities, relationships, and a whole series of them at that. To repeat: don't think, but look!'
> — *Philosophical Inquiries* (1963)

Bibliography

Only sources from which direct quotations are taken in the book's eight main chapters are included in the following list.

Alston, William (1967). 'Religion'. In Paul Edwards (ed.), *The Encyclopedia of Philosophy*, vol. 7: 140–45. New York: Macmillan.

Bourdieu, Perre (1998). *On Television*. New York. The New Press.

Braun, Willi (2000). 'Religion'. In Willi Braun and Russell T. McCutcheon (eds.), *Guide to the Study of Religion*, 3–18. London: Continuum.

Douglas, Mary (1991). *Purity and Danger: An Analysis of the Concepts of Pollution and Taboo*. London: Continuum.

Durkheim, Emile (1995). *The Elementary Forms of Religious Life*. Karen Field (trans.). New York: Free Press.

Fitzgerald, Tim (1996). 'Religion, Philosophy, and Family Resemblances.' *Religion* 26: 215–36.

Freud, Sigmund (1950). *Collected Papers*, vol. 2. London: Hogarth Press.

Gardet, L. (1991). 'Din'. In. B. Lewis and J. Schacht (eds.), *The Encyclopedia of Islam*, vol. 2: 293–96. Leiden: E. J. Brill.

Guthrie, Stewart E. (1993). *Faces in the Clouds: A New Theory of Religion*. New York: Oxford University Press.

Keay, John (2000). *The Great Arc: The Dramatic Tale of How India was Mapped and Everest was Named*. New York: HarperCollins.

Lakoff, George (1987). *Women, Fire, and Dangerous Things*. Chicago: University of Chicago Press.

Lincoln, Bruce (2003). *Holy Terrors: Thinking about Religion after September 11th*. Chicago: University of Chicago Press.

Martin, Luther H. (1987). *Hellenistic Religions: An Introduction*. New York: Oxford University Press.

Penner, Hans (1989). *Impasse and Resolution: A Critique of the Study of Religion*. New York: Peter Lang.

Plato (1993). *The Last Days of Socrates: Euthypro, Apology, Crito, Phaedo*. Hugh Tredennick and Harold Tarrant (trans.), Harold Tarrant (intro.). London: Penguin Books.

Saler, Benson (1993). *Conceptualizing Religion: Immanent Anthropologists, Transcendent Natives, and Unbounded Categories*. Leiden: E. J. Brill.

Smith, Brian K. (1989). *Reflections on Resemblance, Ritual, and Religion*. New York: Oxford University Press.

Smith, Jonathan Z. (1982). *Imagining Religion: From Babylon to Jonestown*. Chicago: University of Chicago Press.

— (1998). 'Religion, Religions, Religious'. In Mark C. Taylor (ed.), *Critical Terms for Religious Studies*, 269–84. Chicago: University of Chicago Press.

— (2000). 'Classification'. In Willi Braun and Russell T. McCutcheon (eds.), *Guide to the Study of Religion*, 35–44. London: Continuum.

Tylor, Edward Burnett (1970). *Religion in Primitive Culture*. Glouchester, MA: Peter Smith Publishing.

Weber, Max (1991). *The Sociology of Religion*. Talcott Parsons (trans.). Boston: Beacon Press.

Williams, Raymond (1976). *Keywords: A Vocabulary of Culture and Society*. New York: Oxford University Press.

Wittgenstein, Ludwig (1968). *Philosophical Investigations*. Oxford: Blackwell.

Resources

The following is a brief list of general resources and scholarly societies, for those wishing to learn more about the academic study of religion.

1. General Works on the Academic Study of Religion

Gregory D. Alles, 'Study of Religion: An Overview'. In *The Encyclopedia of Religion*, 2nd ed., vol. 13, pp. 8760–8767. Macmillan Reference, 2005.

A general essay on the study of religion, the methods and theories used, which is followed by a selection of essays on the study of religion as carried out in various international settings (e.g., Australia, Eastern Europe, Japan, North Africa, North America, South Asia and Sub-Saharan Africa).

Peter Antes, Armin W. Geertz and Randi Warne (eds.), *New Approaches to the Study of Religion*. Vol. 1: Regional, Critical, and Historical Approaches; Vol. 2: Textual, Comparative, Sociological, and Cognitive Approaches. Walter de Gruyter, 2004.

A multi-authored collection of essays on developments in the late twentieth-century study of religion.

Willi Braun and Russell T. McCutcheon (eds.), *Guide to the Study of Religion*. Continuum, 2000.

A multi-authored collection of essays on a variety of topics (e.g., definition, classification, myth, origins, sacred, romaniticism, modernity, etc.), written by some of today's leading scholars.

Walter H. Capps, *Religious Studies: The Making of a Discipline*. Fortress Press, 1995.

A resource by one of the US field's most prominent scholars, comprised of commentaries on many of the field's leading scholars, written in the 'history of ideas' genre.

Peter Connolly, *Approaches to the Study of Religion*. Continuum, 1999.

An introductory text organized around the various methods scholars use to study religion, e.g., anthropological, philosophical, psychology, sociological, etc.

William Deal and Timothy Beal (eds.), *Theory for Religious Studies*. Routledge, 2004.

A brief introduction to many of the leading twentieth-century European theorists of culture, with commentaries on their application to the study of religion and select bibliographies of secondary scholarship on their work.

Mircea Eliade (ed.), *The Encyclopedia of Religion*. Macmillan, 1986 [2nd ed. 2005].

The field's primary multi-volume reference resource, now available in a new and largely supplemented edition.

John Hinnells (ed.), *The Routledge Companion to the Study of Religion*. Routledge, 2005.

A multi-authored collection of essays on a variety of topics (e.g., new religious movements, fundamentalism, religion and science), including essays on the disciplines that comprise the field, written by some of today's leading scholars.

Brian Morris, *Anthropological Studies of Religion: An Introductory Text*. Cambridge University Press, 1998.

Perhaps the best all-around supplementary resource in existence today, especially useful for its discussions of nineteenth- and early- to mid-twentieth-century scholars and methods. A 'must have' for all students of the study of religion.

Malory Nye, *Religion: The Basics*. Routledge, 2003.

An introductory book that focuses on the recent contributions to the study of religion from the field known as culture studies.

J. Samuel Preus, *Explaining Religion: Criticism and Theory from Bodin to Freud*. Yale University Press, 1987.

A modern classic that examines the development of a coherent, naturalistic approach to the study of religion, including chapters on David Hume and Sigmund Freud.

Robert A. Segal (ed.), *The Blackwell Companion to the Study of Religion*. Blackwell, 2006.

A multi-authored collection of twenty-four essays arranged in terms of approaches and topics, written by some of the field's leading thinkers.

Eric J. Sharpe, *Comparative Religion: A History*. Open Court, 1986.

Long considered the standard history of the field, this book is especially useful for nineteenth- and early twentieth-century developments in the field, in Europe and North America.

Jonathan Z. Smith (ed.), *The HarperCollins Dictionary of Religion*. HarperSanFrancisco, 1995.

Perhaps the best single-volume reference resource available today; contains standard dictionary entries along with detailed essays on each of the world's religions as well as on methods and theories.

Mark C. Taylor (ed.), *Critical Terms for Religious Studies*. University of Chicago Press, 1998.

A multi-authored collection of essays on a variety of topics (e.g., belief, body, God, performance, relic), written by some of today's leading scholars.

Jacques Waardenburg (ed.), *Classical Approaches to the Study of Religion*. Vol. 1: Introduction and Anthology; Vol. 2: Bibliography. Mouton, 1973 [Walter de Gruyter, 1999].

A classic two-volume work that provides biographies, excerpts, and extensive bibliographies on the field's nineteenth- and early- to mid-twentieth-century figures.

2. Professional Associations in the Study of Religion

American Academy of Religion (AAR)
http://www.aarweb.org/

The North American field's largest professional association, the AAR, was created in 1963–64 from the formerly named National Association of Biblical Instructors (NABI), and today represents the work of theological, humanistic and social-scientific scholars. Its journal is the *Journal of the American Academy of Religion* and its annual scholarly conference is held each Fall, with each of its various regions holding annual meetings of their own each Spring. Although it has traditionally met concurrently with the annual meeting of the SBL, that arrangement is to end in 2007.

Canadian Corporation for the Study of Religion (CCSR)
http://www.ccsr.ca/

An umbrella organization, or consortium, comprised of seven Canadian scholarly studies that touch upon the study of religion; one of its member associations – the Canadian Society for the Study of Religion – is most relevantly aimed at the wider field of the academic study of religion. The CCSR's bilingual (English/French) journal, *Studies in Religion*, is the primary Canadian periodical in the field.

Council of Societies for the Study of Religion (CSSR)
http://www.cssr.org/

An umbrella organization comprised of several primarily US-based scholarly associations in theological and/or religious studies, which is known mainly through its publications: *Religious Studies Review* (the main book review periodical in the North American field); *CSSR Bulletin* (a professional newsletter); and Directory of Departments of Religious Studies and Theology.

European Association for the Study of Religions (EASR)
http://www.easr.de/

An umbrella organization – which is itself affiliated with the International Association for the History of Religions (IAHR) – comprised of eighteen European professional associations in the academic study of religion, many of which hold their own annual meetings, and some of which publish their own journals.

International Association for the Cognitive Science of Religion
http://www.iacsr.com/

A newly formed scholarly association that promotes the application of cognitive science theories to the cross-cultural study of religion.

International Association for the History of Religions (IAHR)
http://www.iahr.dk/

The primary international umbrella organization in the academic study of religion, comprised of forty-two member associations, from throughout the world, many of which hold their own annual meetings and publish their own journals. The IAHR, whose journal is entitled *Numen*, holds its own Congress, hosted in a different part of the world, once every five years (most recently in Tokyo, Japan, in 2005, and the next in Toronto in 2010).

North American Association for the Study of Religion (NAASR)
http://www.naasr.com/

The North American affiliate of the IAHR, NAASR meets annually at meetings held concurrent with the annual meeting of the AAR; NAASR, which emphasizes theoretical work, publishes *Method & Theory in the Study of Religion*, which is the field's primary journal devoted to theory in the study of religion.

Society of Biblical Literature (SBL)
http://www.sbl-site.org/

The North American field's primary association devoted to historical/critical studies of the Bible. The SBL publishes *Journal of Biblical Studies*. Although it has traditionally met each Fall, concurrent with the annual meeting of the AAR, this arrangement is to end in 2007.

Society for the Scientific Study of Religion (SSSR)
http://www.sssrweb.org/

The North American field's primary association devoted to social-scientific studies of religion (largely emphasizing sociological studies). The SSSR's annual meeting is held early each Fall and its journal is entitled *Journal for the Scientific Study of Religion*.

Index

Bolded page numbers indicate major discussions of the concept or person, as they appear in either the Glossary or Scholars section of the book.